LETTERS

to a

YOUNG SCIENTIST

ALSO BY EDWARD O. WILSON

Why We Are Here: Mobile and the Spirit of a Southern City, with Alex Harris (2012)

The Social Conquest of Earth (2012)

The Leafcutter Ants: Civilization by Instinct, with Bert Hölldobler (2011)

Anthill: A Novel (2010)

Kingdom of Ants: José Celestino Mutis and the Dawn of Natural History in the New World, with José M. Gómez Durán (2010)

The Superorganism: The Beauty, Elegance, and Strangeness of Insect Societies, with Bert Hölldobler (2009)

The Creation: An Appeal to Save Life on Earth (2006)

Nature Revealed: Selected Writings, 1949–2006 (2006)

From So Simple a Beginning: Darwin's Four Great Books, first editions reprinted with introductions (2005)

Pheidole in the New World: A Dominant, Hyperdiverse Ant Genus (2003)

The Future of Life (2002)

Biological Diversity: The Oldest Human Heritage (1999)

Consilience: The Unity of Knowledge (1998)

In Search of Nature (1996)

Journey to the Ants: A Story of Scientific Exploration, with Bert Hölldobler (1994)

Naturalist (1994); new edition (2006)

The Diversity of Life (1992)

The Ants, with Bert Hölldobler (1990); Pulitzer Prize, General Nonfiction, 1991

Success and Dominance in Ecosystems: The Case of the Social Insects (1990)

Biophilia (1984)

Promethean Fire: Reflections on the Origin of the Mind, with Charles J. Lumsden (1983)

Genes, Mind, and Culture: The Coevolutionary Process, with Charles J. Lumsden (1981)

On Human Nature (1978); Pulitzer Prize, General Nonfiction, 1979

Caste and Ecology in the Social Insects, with George F. Oster (1978)

Sociobiology: The New Synthesis (1975); new edition (2000)

A Primer of Population Biology, with William H. Bossert (1971)

The Insect Societies (1971)

The Theory of Island Biogeography, with Robert H. MacArthur (1967)

LETTERS

to a

YOUNG SCIENTIST

Edward O. Wilson

LIVERIGHT PUBLISHING CORPORATION
A Division of W. W. Norton & Company
New York · London

Frontispiece: The author at Gulf Shores, Alabama.
Photograph by Alex Harris.

Printed in the United States of America
First Edition

For information about special discounts for bulk purchases,
please contact W. W. Norton Special Sales
at specialsales@wwnorton.com or 800-233-4830

Manufacturing by Courier Westford
Book design by Abbate Design
Production manager: Anna Oler

ISBN 978-0-87140-377-3

Liveright Publishing Corporation
500 Fifth Avenue, New York, N.Y. 10110
www.wwnorton.com

W. W. Norton & Company Ltd.
Castle House, 75/76 Wells Street, London W1T 3QT

1 2 3 4 5 6 7 8 9 0

To the memory of my mentors,
Ralph L. Chermock and William L. Brown

CONTENTS

PROLOGUE

You Made the Right Choice 13

I · THE PATH TO FOLLOW

1. First Passion, Then Training 21
2. Mathematics 27
3. The Path to Follow 43

II · THE CREATIVE PROCESS

4. What Is Science? 55
5. The Creative Process 69
6. What It Takes 77
7. Most Likely to Succeed 89
8. I Never Changed 95
9. Archetypes of the Scientific Mind 101
10. Scientists as Explorers of the Universe 107

III. A LIFE IN SCIENCE

11. A Mentor and the Start of a Career 119
12. The Grails of Field Biology 127
13. A Celebration of Audacity 143
14. Know Your Subject, Thoroughly 149

IV. THEORY AND THE BIG PICTURE

15. Science as Universal Knowledge 169
16. Searching for New Worlds on Earth 177
17. The Making of Theories 189
18. Biological Theory on a Grand Scale 205
19. Theory in the Real World 219

V. TRUTH AND ETHICS

20. The Scientific Ethic 237

ACKNOWLEDGMENTS 241
PHOTOGRAPH CREDITS 243

LETTERS

to a

YOUNG SCIENTIST

The foraminiferan *Orbulina universa*, a single-celled oceanic organism. Modified from photograph by Howard J. Spero, University of California, Davis.

Prologue

You Made the Right Choice

Dear Friend,

From half a century of teaching students and young professionals in science, I feel privileged and fortunate to have counseled many hundreds of talented and ambitious young people. As a result, I have gleaned a deep knowledge, indeed a philosophy, of what you need to know to succeed in science. I hope you will benefit from the thoughts and stories I will offer you in the letters to follow.

First and foremost, I urge you to stay on the path you've chosen, and to travel on it as far as you can. The world needs you—badly. Humanity is now fully in the technoscientific age, and there is no turning back. Although its rate of increase varies among its many disciplines, scientific knowledge doubles every fifteen to twenty years. And so it has been since the 1600s, achieving a prodigious magnitude today. And like all unfettered exponential growth given enough time, it seems decade by decade to be ascending almost vertically. High technology runs at comparable pace alongside it. Science and technology, bound in tight symbiotic alliance, pervade every dimension of our lives. They hide no long-lasting secrets. They are open to everyone, everywhere. The Internet and all the other accouterments of digital technology have rendered communication global and instant. Soon all published knowledge in both science and the humanities will be available with a few keystrokes.

In case this assessment seems a bit feverish (although I suspect it is not, really), I'll provide an example of a quantum leap in which I was fortunate to play a role. It occurred in taxonomy, the classification of organisms, until recently a notoriously old-fashioned and sluggish discipline. Back in 1735, Carl Linnaeus, a Swedish naturalist who ranked with Isaac Newton as the best-known scientist of the eighteenth century, launched one of the most audacious research projects of all time. He proposed to discover and classify every kind of plant and animal on Earth. In 1759, to streamline the process, he began to give each species a double Latinized name, such as *Canis familiaris* for the domestic dog and *Acer rubrum* for the American red maple.

Linnaeus had no idea, not even to the power of 10 (that is, whether 10,000, or 100,000, or 1,000,000), of the magnitude of his self-assigned task. He guessed that plant species, his specialty, would turn out to number around 10,000. The richness of the tropical regions were unknown to him. The number of known and classified plant species today is 310,000 and is expected to reach 350,000. When animals and fungi are added, the total number of species currently known is in excess of 1.9 million—and is expected to eventually reach 10 million or more. Of bacteria, the "dark matter" of living diversity, only about 10,000 kinds are currently known (in 2013), but the number is accelerating and is likely to add millions of species to the global roster. So, just as in Linnaeus's time 250 years ago, most of life on Earth remains unknown.

The still-deep pit of ignorance about biodiversity is a problem not just for specialists but for everyone. How are we all going to manage the planet and keep it sustainable if we know so little about it?

Until recently, the solution seemed out of reach. Hardworking scientists have been able to add only about eighteen thousand new species each year. If this rate were to continue, it would take two centuries or longer to account for all of Earth's biodiversity, a period nearly as long as that from the Linnaean initiative to the present time. What is the reason for this bottleneck? Until recently the problem was one of technology, and it appeared insoluble. For historical reasons, the great bulk of reference specimens and printed literature about them was confined to a relatively small number of museums, located in a few cities in Western Europe and North America. To conduct basic research on taxonomy, it was often necessary to visit these distant places. The only alternative was to arrange to have the specimens and literature mailed, always a time-consuming and risky operation.

By the turn of the twenty-first century, biologists were looking for a technology that could somehow solve the problem. In 2003 I suggested what in retrospect seems the obvious solution: the creation of the online Encyclopedia of Life, which would include digitized, high-resolution photographs of reference specimens, with all information on each species, updated continuously. It was to be an open source, with new entries screened by "curators" expert in each group of species, such as centipedes, bark beetles, and conifers. The project was funded by 2005, and with the parallel Census of Marine Life, it has accelerated taxonomy, as well as those branches of biology dependent on accurate classification. At the time I write, over half the known species on Earth have been incorporated. The knowledge is available to anyone, anytime, anywhere, for free, at a keystroke (EOL.com).

So swift do advances like this in biodiversity studies occur, so startling the twists and turns in every discipline, the future of the technoscientific revolution cannot be assayed for any branch of science even just a decade ahead. Of course, there will come a time when the exponential growth in discovery and cumulative knowledge must peak and level off. But that won't matter to you. The revolution will continue for at least most of the twenty-first century, during which it will render the human condition radically different from what it is today. Traditional disciplines of research will metamorphose, by today's standards, into barely recognizable forms. In the process they will spin off new fields of research—science-based technology, technology-based science, and industry based on technology and science. Eventually all of science will coalesce into a continuum of description and explanation through which any educated person can travel by guidelines of principles and laws.

The introduction to science and scientific careers that I will give you in this series of letters is not traditional in form or tone. I mean it to be as personal as possible, using my experiences in research and teaching to provide a realistic image of the challenges and rewards you can expect as you pass through a life in science.

I

THE PATH *to* FOLLOW

Merit badge symbol for "Zoology" in 1940. *Boy Scout Handbook*, Boy Scouts of America, fourth edition (1940).

One

First Passion, Then Training

I BELIEVE IT WILL HELP for me to start with this letter by telling you who I really am. This requires your going back with me to the summer of 1943, in the midst of the Second World War. I had just turned fourteen, and my hometown, the little city of Mobile, Alabama, had been largely taken over by the buildup of a wartime shipbuilding industry and military air base. Although I rode my bicycle around the streets of Mobile a couple of times as a potential emergency messenger, I remained oblivious to the great events occurring in the city and world. Instead, I spent a lot of my spare time—not required to be at school—earning merit badges in my quest to reach the Eagle rank in the Boy Scouts of America. Mostly, however, I explored nearby swamps and forests, collecting ants and butterflies. At home I attended to my menagerie of snakes and black widow spiders.

Global war meant that very few young men were available to serve as counselors at nearby Boy Scout Camp Pushmataha. The recruiters, having heard of my extracurricular activities, had asked me, I assume in desperation, to serve as the nature counselor. I was, of course, delighted with the prospect of a free summer camp experience doing approximately what I most wanted to do anyway. But I arrived at Pushmataha woefully underaged and underprepared in much of anything but ants and butterflies. I was nervous. Would the other scouts, some older than I, laugh at what I had to offer? Then I had an inspiration: *snakes*. Most people are simultaneously frightened, riveted, and instinctively interested in snakes. It's in the genes. I didn't realize it at the time, but the south-central Gulf coast is home to the largest variety of snakes in North America, upward of forty species. So upon arrival I got some of the other campers to help me build some cages from wooden crates and window screen. Then I directed all residents of the camp to join me in a summer-long hunt for snakes whenever their regular schedules allowed.

Thereafter, on an average of several times a day, the cry rang out from somewhere in the woods: Snake! Snake! All within hearing distance would rush to the spot, calling to others, while I, snake-wrangler-in-chief, was fetched.

If nonvenomous, I would simply grab it. If

venomous, I would first press it down just behind the head with a stick, roll the stick forward until its head was immobile, then grasp it by the neck and lift it up. I'd then identify it for the gathering circle of scouts and deliver what little I knew about the species (usually very little, but they knew less). Then we would walk to headquarters and deposit it in a cage for a residence of a week or so. I'd deliver short talks at our zoo, throw in something new I learned about local insects and other animals. (I scored zero on plants.) The summer rolled by pleasantly for me and my small army.

The only thing that could interrupt this happy career was, of course, a snake. I have since learned that all snake specialists, scientists and amateurs alike, apparently get bitten at least once by a venomous snake. I was not to be an exception. Halfway through the summer I was cleaning out a cage that contained several pygmy rattlesnakes, a venomous but not deadly species. One coiled closer to my hand than I'd realized, suddenly uncoiled, and struck me on the left index finger. After first aid in a doctor's office near the camp, which was too late to do any good, I was sent home to rest my swollen left hand and arm. Upon returning to Pushmataha a week later, I was instructed by the adult director of the camp, as I already had been by my parents, that I was to catch no more venomous snakes.

At the end of the season, as we all prepared to leave, the director held a popularity poll. The campers, most of whom were assistant snake hunters, placed me second, just behind the chief counselor. I had found my life's work. Although the goal was not yet clearly defined then in my adolescent mind, I was going to be a scientist—and a professor.

Through high school I paid very little attention to my classes. Thanks to the relatively relaxed school systems of south Alabama in wartime, with overworked and distracted teachers, I got away with it. One memorable day at Mobile's Murphy High School, I captured with a sweep of my hand and killed twenty houseflies, then lined them up on my desk for the next hour's class to find. The following day the teacher, a young lady with considerable aplomb, congratulated me but kept a closer eye on me thereafter. That is all I remember, I am embarrassed to say, about my first year in high school.

I arrived at the University of Alabama shortly after my seventeenth birthday, the first member of my family on either side to attend college. I had by this time shifted from snakes and flies to ants. Now determined to be an entomologist and work in the outdoors as much as possible, I kept up enough effort to make A's. I found that not very difficult (it is, I'm told, *very* different today), but soaked up all the elementary and intermediate chemistry and biology available.

Harvard University was similarly tolerant when I arrived as a Ph.D. student in 1951. I was considered a prodigy in field biology and entomology, and was allowed to make up the many gaps in general biology left from my happy days in Alabama. The momentum I built up in my southern childhood and at Harvard carried through to an appointment at Harvard as assistant professor. There followed more than six decades of fruitful work at this great university.

I've told you my Pushmataha-to-Harvard story not to recommend my kind of eccentricity (although in the right circumstances it could be of advantage); and I disavow my casual approach to early formal education. I grew up in a different age. You, in contrast, are well into a different era, where opportunity is broader but more demanding.

My confessional instead is intended to illustrate an important principle I've seen unfold in the careers of many successful scientists. It is quite simple: put passion ahead of training. Feel out in any way you can what you most want to do in science, or technology, or some other science-related profession. Obey that passion as long as it lasts. Feed it with the knowledge the mind needs to grow. Sample other subjects, acquire a general education in science, and be smart enough to switch to a greater love if one appears. But don't just drift through courses in science hoping that love will come to you. Maybe it will, but don't take the chance. As in other big choices in your life, there is too much at stake. Decision and hard work based on enduring passion will never fail you.

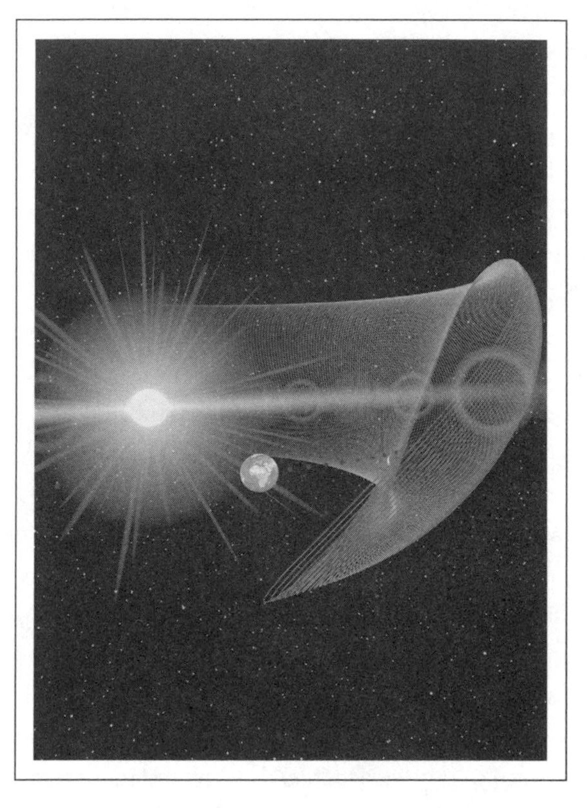

Reconstructed path of the "Trojan" asteroid 2010 TK_7, during 165 years, seen from outside Earth's orbit. Modified from drawing. © Paul Wiegert, University of Western Ontario.

Two

MATHEMATICS

LET ME MOVE ON quickly, and before everything else remaining, to a subject that is both a vital asset for and a potential barrier to your career: mathematics, the great bugbear for many would-be scientists. I mention this not to nag but to encourage and help. I mean in this letter to put you at ease. If you're already well prepared—let us say you've picked up calculus and analytic geometry—if you like to solve puzzles, and if you think logarithms are a neat way to express variables across orders of magnitude, then good for you; your capability is a comfort to me. I won't worry so much about you, at least not right away. But keep in mind that a strong mathematical background does not—I repeat, does not—guarantee success in science. I will return to this caveat later, so please stay focused. Actually, I have a lot more to say to math lovers in particular.

If, on the other hand, you are a bit short in mathematical training, even very short, relax. You are far from alone in the community of scientists, and here is a professional secret to encourage you: many of the most successful scientists in the world today are mathematically no more than semiliterate. A metaphor will clarify the paradox in this statement. Where elite mathematicians often serve as architects of theory in the expanding realm of science, the remaining large majority of basic and applied scientists map the terrain, scout the frontier, cut the pathways, and raise the first buildings along the way. They define the problems that mathematicians, on occasion, may help solve. They think primarily in images and facts, and only marginally in mathematics.

You may think me foolhardy, but it's been my habit to brush aside the fear of mathematics when talking to candidate scientists. During my decades of teaching biology at Harvard, I watched sadly as bright undergraduates turned away from the possibility of a scientific career, or even from nonrequired courses in the sciences, because they were afraid of failure in the math that might be required. Why should I care? Because such math-phobes deprive science of an immeasurable amount of sorely needed talent and deprive the many scientific disciplines of some of their most creative young people. This is a hemorrhage of brainpower we need to stanch.

Now I will tell you how to ease your anxieties. Understand that mathematics is a language, ruled like verbal languages by its own grammar and system of logic. Any person with average quantitative intelligence who learns to read and write mathematics at an elementary level will have little difficulty understanding math-speak.

Let me give you an example of the interplay of visual images and simple mathematical statements. I've chosen to reveal the undergirding of two relatively advanced disciplines in biology: population genetics and population ecology.

Consider this interesting fact. You have (or had) 2 parents, 4 grandparents, 8 great-grandparents, and 16 great-great-grandparents. In other words, since each person has to have two parents, the number of your direct forebears doubles every generation. The mathematical summary is $N = 2^x$. The parameter N is the number of a person's ancestors x generations back in time. How many of your ancestors existed 10 generations ago? We don't have to write out each generation in turn. Instead you can use $N = 2^x = 2^{10}$, or, put the other way, $2^{10} = N$. So the answer is when $x = 10$ generations, you have $N = 1{,}024$ ancestors. Now reverse the timeline to forward and ask how many descendants you can expect to have 10 generations from now. The whole thing gets much more complicated in the case of descendants—we don't really know how many children we will have—but to state the basic idea, it is all right to specify, in a way mathematicians often do, that each couple will have two surviving children and the length of the generations will be constant from one generation to the next. (Two children on average is not far from the actual rate in the United States today, and is close to the number 2.1, or 21 children for every 100 couples, needed to maintain a constant population size of native-born.) Then in 10 generations you will have 1,024 descendants.

What are we to make of this? For one thing, it is a humbling picture of the origin and the fate of one person's genes. The fact is that sexual reproduction chops apart the combinations that prescribe each person's traits and recombines half of them with somebody else's genes to make the next generation. Over a very few generations, each parent's combination will be dissolved in the gene pool of the population as a whole. Suppose you have a distinguished forebear who fought in the American Revolution, during which another roughly 250 of your other direct ancestors lived, including possibly a horse thief or two or three. (One of my 8 great-great-grandfathers, a confederate veteran of the Civil War, was a notoriously tricky horse trader, if not quite a thief.)

Mathematicians like to take the measurement of exponential growth from just counting jumps from one generation to the next, to the much more general state to fit a large population over a particular moment in time (to the hour, minute, or shorter interval as they choose). This is done with calculus, which expresses the growth of population in the form $dN/dt = rN$, which says in any very short interval of time, dt, the population is growing a certain amount, dN, and the rate is the differential dN/dt. In the case of exponential growth, N, the number of individuals in the population at the instant is multiplied by r, a constant that depends on the nature of the population and the circumstances in which it lives.

You can pick any N and r that interests you, and run with these two parameters for as long as you choose. If the differential dN/dt is larger than zero and the population (say, of bacteria or mice or humans) is allowed in theory to increase at the same rate indefinitely, in a surprising few years the population would weigh more than Earth, than the solar system, and finally than the entire known universe.

It is easy to produce fantastical results with mathematically correct theory. There are a lot of models that fit reality and produce factual implications that can jolt us into a new way of thinking. A famous one learned from exponential growth of the kind I've just described is the following. Suppose there is a pond, and a lily pad is put in the pond. This first pad doubles into two pads, each of which also doubles. The pond will fill and no more pads can double at the end of thirty days. When is the pond half full? On the twenty-ninth day. This elementary bit of mathematics, obvious upon commonsense reflection, is one of many ways to emphasize the risks of excessive population growth. For two centuries the global human population has been doubling every several generations. Most demographers and economists agree that a global population of more than ten billion would make it very difficult to sustain the planet. We recently shot past seven billion. When was the Earth half full? Decades ago, say the experts. Humanity is racing toward the wall.

The longer you wait to become at least semiliterate in math, the harder the language of mathematics will be to master—again the same as in verbal languages. But it can be done, and at any age. I speak as an authority on this subject, because I am an extreme case. Having spent my pre-college years in relatively poor southern schools, I didn't take algebra until my freshman year at the University of Alabama. My student days being at the end of the Depression, algebra just wasn't offered. I finally got around to calculus as a thirty-two-year-old tenured professor at Harvard, where I sat uncomfortably in classes with undergraduate students only a bit more than half my age. A couple of them were students in a course on evolutionary biology I was teaching. I swallowed my pride and learned calculus.

Admittedly, I was never more than a C student while catching up, but I was reassured somewhat by the discovery that superior mathematical ability is similar to fluency in foreign languages. I might have become fluent with more effort and sessions talking with the natives, but, being swept up with field and laboratory research, I advanced only by a small amount.

A true gift in mathematics is probably hereditary in part. What this means is that variation within a group in ability is due in some measurable degree to differences in genes among the group members rather than entirely just to the environment in which they grew up. There is nothing that you and I can do about hereditary differences, but it is possible to greatly reduce the part of the variation due to the environment simply by raising our ability through education and practice. Mathematics is convenient in that it can be achieved by self-instruction.

Having gone this far, I believe I should go on a bit further, and explain how fluency is achieved by those who wish to attain it. Practice allows elementary operations (such as, "If $y = x + 2$, then $x = y - 2$") to be effortlessly retrieved in memory, much like words and phrases (such as "effortlessly retrieved in memory"). Then, in the way verbal phrases are almost unconsciously put together in sentences and sentences are built into paragraphs, mathematical operations can be put together with ease in ever more complex sequences and structures. There is, of course, much more to mathematical reasoning. There are, for example, the positioning and proving of theorems, the exploration of series, and the invention of new modes of geometry. But short of these adventures of advanced pure mathematics, the language of mathematicians can be learned well enough to understand the majority of mathematical statements made in scientific publications.

Exceptional mathematical fluency is required in only a few disciplines. Particle physics, astrophysics, and information theory come to mind. Far more important throughout the rest of science and its applications, however, is the ability to form concepts, during which the researcher conjures images and processes in visual images by intuition. It's something everyone already does to some degree.

In your imagination, be the great eighteenth century physicist Isaac Newton. Think of an object falling through space. (In the legend, he was attracted to an apple falling from the tree to the ground.) Make it from high up, like a package dropped from an airplane. The object accelerates to about 120 miles an hour, then holds that velocity until it hits the ground. How can you account for this acceleration up to but not beyond terminal velocity? By Newton's laws of motion, plus the existence of air pressure, the kind used to propel a sailboat.

Stay as Newton a moment longer. Notice as he did how light passing through curved glass sometimes comes out as a rainbow of colors, always ranging from red to yellow to green to blue to violet. Newton thought that white light is just a mix of the colored lights. He proved it by passing the same array of colors back through a prism, turning the mix back into white light. Scientists were later to understand, from other experiments and mathematics, that the colors are radiations differing in wavelength. The longest we are able to see creates the sensation of red, and the shortest the sensation of blue.

You likely knew all that already. Whether you did or not, let's go on to Darwin. As a young man in the 1830s, he made a five-year voyage on a British government vessel, the HMS *Beagle*, around the coast of South America. He took that long period to explore and think broadly and deeply about the natural world. He found, for example, a lot of fossils, some of extinct large animals similar to modern species like horses, tigers, and rhinoceroses—yet different in many important ways than these modern equivalents. Were they just victims of the biblical flood that Noah failed to save? But that couldn't be, Darwin must have realized; Noah saved all the kinds of animals. The South American species were obviously not among them.

As the young naturalist went from one part of the continent to another, he noticed something else: some kinds of living birds and other animals found in one locality were replaced by closely similar yet distinctly different kinds in another. What, he must have thought, is going on here? Today we know it was evolution, but that answer was not open to the young man. Anything that so openly contradicted holy scripture was considered heresy back home in England, and Darwin had trained for the ministry at the University of Cambridge.

When he finally accepted evolution, during the voyage back home, he soon began puzzling over the *cause* of evolution. Was it divine guidance? Not likely. The inheritance of changes caused directly by the environment, as suggested earlier by the French zoologist Jean-Baptiste Lamarck? Others had already rejected that theory. How about progressive change built into the heredity of organisms that unfolds from one generation to the next? That was hard to imagine, and in any case Darwin was soon figuring out another process, natural selection, in which varieties within a species—varieties that survive longer, reproduce more, or both—replace other, less successful varieties in the same species.

The idea and its supporting logic came in pieces to Darwin while walking around his rural home, riding in a carriage, or, in one important case, sitting in his garden staring at an anthill. Darwin admitted later that if he couldn't explain how sterile ant workers passed on their worker anatomy and behavior to later generations of sterile ant workers, he might have to abandon his theory of evolution. He conceived the following solution: the worker traits are passed on through the mother queen; workers have the same heredity as the queen, but are reared in a different, stultifying environment. One day, during this lucubration, when a housemaid saw him staring at an anthill in the garden, she made reference to a famous prolific novelist living nearby when she said (it is reported), "What a pity Mr. Darwin doesn't have a way to pass his time, like Mr. Thackeray."

Everyone sometimes daydreams like a scientist at one level or another. Ramped up and disciplined, fantasies are the fountainhead of all creative thinking. Newton dreamed, Darwin dreamed, you dream. The images evoked are at first vague. They may shift in form and fade in and out. They grow a bit firmer when sketched as diagrams on pads of paper, and they take on life as real examples are sought and found.

Pioneers in science only rarely make discoveries by extracting ideas from pure mathematics. Most of the stereotypical photographs of scientists studying rows of equations written on blackboards are instructors explaining discoveries already made. Real progress comes in the field writing notes, at the office amid a litter of doodled paper, in the corridor struggling to explain something to a friend, at lunchtime, eating alone, or in a garden while walking. To have a eureka moment requires hard work. And focus. A distinguished researcher once commented to me that a real scientist is someone who can think about a subject while talking to his or her spouse about something else.

Ideas in science emerge most readily when some part of the world is studied for its own sake. They follow from thorough, well-organized knowledge of all that is known or can be imagined of real entities and processes within that fragment of existence. When something new is encountered, the follow-up steps will usually require the use of mathematical and statistical methods in order to move its analysis forward. If that step proves technically too difficult for the person who made the discovery, a mathematician or statistician can be added as a collaborator. As a researcher who has coauthored many papers with mathematicians and statisticians, I offer the following principle with confidence. Let's call it Principle Number One:

> It is far easier for scientists to acquire needed collaboration from mathematicians and statisticians than it is for mathematicians and statisticians to find scientists able to make use of their equations.

For example, when I sat down in the late 1970s with the mathematical theorist George Oster to work out the principles of caste and division of labor in the social insects, I supplied the details of what had been discovered in nature and in the laboratory. Oster then drew methods from his diverse toolkit to create theorems and hypotheses concerning this real world laid before him. Without such information Oster might have developed a general theory in abstract terms that covers all possible permutations of castes and division of labor in the universe, but there would have been no way of deducing which ones of these multitude options exist on Earth.

This imbalance in the role of observation and mathematics is especially the case in biology, where factors in a real-life phenomenon are often either misunderstood or never noticed in the first place.

The annals of theoretical biology are clogged with mathematical models that either can be safely ignored or, that when tested, fail. Possibly no more than 10 percent have any lasting value. Only those linked solidly to knowledge of real living systems have much chance of being used.

If your level of mathematical competence is low, plan on raising it, but meanwhile know that you can do outstanding work with what you have. Such is markedly true in fields built largely upon the amassing of data, including, for example, taxonomy, ecology, biogeography, geology, and archaeology. At the same time, think twice about specializing in fields that require a close alternation of experiment and quantitative analysis. These include the greater part of physics and chemistry, as well as a few specialties within molecular biology. Learn the basics of improving your mathematical literacy as you go along, but if you remain weak in mathematics, seek happiness elsewhere among the vast array of scientific specialties. Conversely, if tinkering and mathematical analysis give you joy, but not the amassing of data for their own sake, stay away from taxonomy and the other more descriptive disciplines just listed.

Newton, for example, invented calculus in order to give substance to his imagination. Darwin by his own admission had little or no mathematical ability, but was able with masses of information he had accumulated to conceive a process to which mathematics was later applied. An important step for you to take is to find a subject congenial to your level of mathematical competence that also interests you deeply, and focus on it. In so doing, keep in mind Principle Number Two:

> For every scientist, whether researcher, technologist, or teacher, of whatever competence in mathematics, there exists a discipline in science for which that level of mathematical competence is enough to achieve excellence.

A relativistic jet formed as gas and stars fall into a black hole; artist's conception. Modified from painting by Dana Berry of the Space Telescope Science Institute (STScI). http://hubblesite.org/newscenter/archive/releases/1990/29/image/a/warn/.

Three

The Path to Follow

THE PURPOSE OF THIS LETTER is to help orient you among your colleagues.

When I was a sixteen-year-old senior in high school, I decided the time had come to choose a group of animals on which to specialize when I entered college the coming fall. I thought about spear-winged flies of the taxonomic family Dolichopodidae, whose tiny bodies sparkle like animated gemstones in the sun. But I couldn't get the right equipment or literature to study them. So I turned to ants. By sheer luck, it was the right choice.

Arriving at the University of Alabama at Tuscaloosa, with my well-prepared and identified beginner's collection of ants, I reported to the biology faculty to begin my freshman year of research. Perhaps charmed by my naïveté, or perhaps recognizing an embryonic academic when they saw one, or both, I was welcomed by the faculty and given a stage microscope and personal laboratory space. This support, on top of my earlier success as nature counselor at Camp Pushmataha, buoyed my confidence that I had the right subject and the right university.

My good fortune came from an entirely different source, however. It was choosing ants in the first place. These little six-legged warriors are the most abundant of all insects. As such, they play major roles in land environments around the world. Of equal importance for science, ants, along with termites and honeybees, have the most advanced social systems of all animals. Yet, surprisingly, at the time I entered college only about a dozen scientists around the world were engaged full-time in the study of ants. I had struck gold before the rush began. Almost every research project I began thereafter, no matter how unsophisticated (and all were unsophisticated), yielded discoveries publishable in scientific journals.

What does my story mean to you? A great deal. I believe that other experienced scientists would agree with me that when you are selecting a domain of knowledge in which to conduct original research, it is wise to look for one that is sparsely inhabited. Judge opportunity by how few there are of other students and researchers in one field versus another. This is not to deny the essential requirement of broad training, or the value of apprenticing yourself to researchers and programs of high quality. Or that it also helps to make a lot of friends and colleagues of your age in science for mutual support.

Nonetheless, through it all, I advise you to look for a chance to break away, to find a subject you can make your own. That is where the quickest advances are likely to occur, as measured by discoveries per investigator per year. Therein you have the best chance to become a leader and, as time passes, to gain growing freedom to set your own course.

If a subject is already receiving a great deal of attention, if it has a glamorous aura, if its practitioners are prizewinners who receive large grants, stay away from that subject. Listen to the news coming from the current hubbub, learn how and why the subject became prominent, but in making your own long-term plans be aware it is already crowded with talented people. You would be a newcomer, a private amid bemedaled first sergeants and generals. Take a subject instead that interests you and looks promising, and where established experts are not yet conspicuously competing with one another, where few if any prizes and academy memberships have been given, and where the annals of research are not yet layered with superfluous data and mathematical models. You may feel lonely and insecure in your first endeavors, but, all other things being equal, your best chance to make your mark and to experience the thrill of discovery will be there.

You may have heard the military rule for the summoning of troops to a battlefield: "March to the sound of the guns." In science the opposite is the one for you, as expressed in Principle Number Three:

> March away from the sound of the guns. Observe the fray from a distance, and while you are at it, consider making your own fray.

Once you have settled upon a subject you can love, your potential to succeed will be greatly enhanced if you study it enough to become a world-class expert. This goal is not as difficult as it may seem, even for a graduate student. It is not overly ambitious. There are thousands of subjects in science, sprinkled through physics and chemistry, biology and the social sciences, where it is possible in a short time to attain the status of an authority. If the subject is still thinly populated, you can with diligence and hard work even become *the* world authority at a young age. Society needs this level of expertise, and it rewards the kind of people willing to acquire it.

The already existing information, and what you yourself will discover, may at first be skimpy and difficult to connect to other bodies of knowledge. If this proves to be the case, that's very good. Why should the path to a scientific frontier usually be hard rather than easy? The answer is stated as Principle Number Four:

> In the search for scientific discoveries, every problem is an opportunity. The more difficult the problem, the greater the likely importance of its solution.

The truth of this guidebook dictum can be most clearly seen in extreme cases. The sequencing of the human genome, the search for life on Mars, and the finding of the Higgs boson were each of profound importance for medicine, biology, and physics, respectively. Each required the work of thousands, and cost billions. Each was worth all the trouble and expense. But on a far smaller scale, in fields and subjects less advanced, a small squad of researchers, even a single individual, can with effort devise an important experiment at relatively low cost.

This brings me to the ways in which scientific problems are found and discoveries made. Scientists, mathematicians among them, follow one of two pathways. First, early in the research a problem is identified, and then a solution is sought. The problem may be relatively small (for example, what is the average life span of a Nile crocodile?) or large (what is the role of dark matter in the universe?). As an answer emerges, other phenomena are typically discovered, and other questions raised. The second strategy is to study a subject broadly, while searching for any previously unknown or even unimagined phenomena. The two strategies of original scientific research are stated as Principle Number Five:

> For every problem in a given discipline of science, there exists a species or other entity or phenomenon ideal for its solution. (Example: a kind of mollusk, the sea hare *Aplysia*, proved ideal for exploring the cellular base of memory.)
>
> Conversely, for every species or other entity or phenomenon, there exist important problems for the solution of which it is ideally suited. (Example: bats were logical for the discovery of sonar.)

Obviously, both strategies can be followed, together or in sequence, but by and large scientists who use the first strategy are instinctive problem solvers. They are prone by taste and talent to select a particular kind of organism, or chemical compound, or elementary particle, or physical process, to answer questions about its properties and roles in nature. Such is the predominant research activity in the physical sciences and molecular biology.

The following example is a fictitious scenario of

the first strategy, but, I promise you, is close to true dramas that occur in laboratories:

> *Think of a small group of white-coated men and women in a laboratory—early one afternoon, let us say—watching the readout on a digital monitor. That morning, before setting up the experiment, they were in a nearby conference room, conferring, occasionally taking turns at the blackboard to make an argument. With coffee break, lunch, a few jokes, they decide to try this or that. If the data in the readout are as expected, that will be very interesting, a real lead. "It would be what we're looking for," the team leader says. And it is! The object of the search is the role of a new hormone in the mammalian body. First, though, the team leader says, "Let's break out some champagne. Tonight, we'll all have dinner at a decent restaurant and start talking about what comes next."*

In biology, the first, problem-oriented strategy (for every problem, an ideal organism) has resulted in a heavy emphasis on several dozen "model species." When in your studies you take up the molecular basis of heredity you will learn a great deal that came from a bacterium living in the human gut, *E. coli* (condensed from its full scientific name, *Escherichia coli*). For the organization of cells in the nervous system, there is inspiration from the roundworm

C. elegans (*Caenorhabditis elegans*). And for genetics and embryonic development, you will become familiar with fruitflies of the iconic genus *Drosophila*. This is, of course, as it should be. Better to know one thing in depth rather than a dozen things at their surface only.

Still, keep in mind that during the next few decades there will be at most only a few hundred model species, out of close to two million other species known to science by scarcely more than a brief diagnosis and a Latinized name. Although the latter multitude tend to possess most of the same basic processes discovered in the model species, they further display among them an immense array of idiosyncratic traits in anatomy, physiology, and behavior. Think, in one sweep of your mind, first of a smallpox virus, then of all you know about it. Then the same for an amoeba, and then on to a maple tree, blue whale, monarch butterfly, tiger shark, and human being. The point is that each such species is a world unto itself, with a unique biology and place in an ecosystem, and, not least, an evolutionary history thousands to millions of years old.

When a biologist studies a group of species, ranging anywhere from, say, elephants with three living species to ants with fourteen thousand species, he or she typically aims to learn everything possible over a wide range of biological phenomena. Most researchers working this way, following the second strategy of research, are properly called scientific naturalists. They love the organisms they study for their own sake. They enjoy studying creatures in the field, under natural conditions. They will tell you, correctly, that there is infinite detail and beauty even in those that people at first find least attractive—slime molds, for example, dung beetles, cobweb spiders, and pit vipers. Their joy is in finding something new, the more surprising the better. They are the ecologists, taxonomists, and biogeographers. Here is a scenario of a kind I have personally experienced many times:

> *Think of two biologists hunting in a rain forest, packing heavy collecting equipment, with an online field guide waiting back at camp and DNA analysis at the home laboratory. "Good God, what is that?" one says, pointing to a small, strangely shaped, brilliantly colored animal plastered onto the underside of a palm leaf. "I think it's a hylid frog," his companion replies. "No, no, wait, I've never seen anything like it. It's got to be something new. What the hell is it? Listen, get close, and be careful, don't lose it. There, got it. We're not going to preserve it yet. You never know, it might be an endangered species. Let's take it back alive to camp and see what we can find on the Encyclopedia of Life website. There's that guy at Cornell, he knows all the amphibians like*

this one pretty well, I think. We might check in with him. First, though, we ought to look around for more specimens, get all the information we can." The pair arrive back at camp and start pulling up information. What they find is astonishing. The frog appears to be a new genus, unrelated to any other previously known. Scarcely believing, the pair go online to spread news of the discovery to other specialists around the world.

The potential paths you can follow with a scientific career are vast in number. Your choice may take you into one of the scenarios I've described, or not. The subject for you, as in any true love, is one in which you are interested and that stirs passion and promises pleasure from a lifetime of devotion.

II

THE CREATIVE PROCESS

Charles Darwin at 31 years of age. Modified from painting by George Richmond.

Four

WHAT IS SCIENCE?

WHAT IS THIS grand enterprise called *science* that has lit up heaven and earth and empowered humanity? It is organized, testable knowledge of the real world, of everything around us as well as ourselves, as opposed to the endlessly varied beliefs people hold from myth and superstition. It is the combination of physical and mental operations that have become increasingly the habit of educated peoples, a culture of illuminations dedicated to the most effective way ever conceived of acquiring factual knowledge.

You will have heard the words "fact," "hypothesis," and "theory" used constantly in the conduct of scientific research. When separated from experience and spoken of as abstract ideas they are easily misunderstood and misapplied. Only in case histories of research, by others and soon by you, will their full meaning become clear.

I'll give you an example of my own to show you what I mean. I started with a simple observation: ants remove their dead from the nests. Those of some species just dump the corpses at random outside, while those of other species place them on piles of refuse that might be called "cemeteries." The problem I saw in this behavior was simple but interesting: How does an ant know when another ant is dead? It was obvious to me that the recognition was not by sight. Ants recognize a corpse even in the complete darkness of the underground nest chambers. Furthermore, when the body is fresh and in a lighted area, and even when it is lying on its back with its legs in the air, others ignore it. Only after a day or two of decomposition does a body become a corpse to another ant. I guessed (made a hypothesis) that the undertaker ants were using the odor of decomposition to recognize death. I further thought it likely (second hypothesis) that their response was triggered by only a few of the substances exuded from the body of the corpse. The inspiration for the second hypothesis was an established principle of evolution: animals with small brains, which are the vast majority of animals on Earth, tend to use the simplest set of available cues to guide them through life. A dead body offers dozens or hundreds of chemical cues from which to choose.

Human beings can sort out these components. But ants, with brains one-millionth the size of our own, cannot.

So if the hypotheses are true, which of these substances might trigger the undertaker response—all of them, a few of them, or none? From chemical suppliers I obtained pure synthetic samples of various decomposition substances, including skatole, the essence of feces; trimethylamine, the dominant odor of rotting fish; and various fatty acids and their esters of a kind found in dead insects. For a while my laboratory smelled like a combination of charnel house and sewer. I put minute amounts on dummy ant corpses made of paper and inserted them into ant colonies. After a lot of smelly trial and error I found that oleic acid and one of its oleates trigger the response. The other substances were either ignored or caused alarm.

To repeat the experiment another way (and admittedly for my and others' amusement), I dabbed tiny amounts of oleic acid on the bodies of living worker ants. Would they become the living dead? Sure enough, they did become zombies, at least broadly defined. They were picked up by nestmates, their legs kicking, carried to the cemetery, and dumped. After they had cleaned themselves awhile, they were permitted to rejoin the colony.

I then came up with another idea: insects of all kinds that scavenge for a living, such as blowflies and scarab beetles, find their way to dead animals or dung by homing in on the scent. And they do so by using a very small number of the decomposition chemicals present. A generalization of this kind, widely applied, with at least a few facts here and there and some logical reasoning behind it, is a theory. Many more experiments, applied to other species, would be required to turn it into what can be confidently called a fact.

What, then, in broadest terms is the scientific method? The method starts with the discovery of a phenomenon, such as a mysterious ant behavior, or a previously unknown class of organic compounds, or a newly discovered genus of plants, or a mysterious water current in the ocean's abyss. The scientist asks: What is the full nature of this phenomenon? What are its causes, its origin, its consequence? Each of these queries poses a problem within the ambit of science. How do scientists proceed to find solutions? Always there are clues, and opinions are quickly formed from them concerning the solutions. These opinions, or just logical guesses as they often are, are the hypotheses. It is wise at the outset to figure out as many different solutions as seem possible, then test the whole, either one at a time or in bunches, eliminating all but one. This is called the method of multiple competing hypotheses. If something like this analysis is not followed—and, frankly, it often is not—individual scientists tend to fixate on one alternative or another, especially if they authored it. After all, scientists are human.

Only rarely does an initial investigation result in a clear delineation of all possible competing hypotheses. This is especially the case in biology, in which multiple factors are the rule. Some factors remain undiscovered, and those that have been discovered commonly overlap and interact with one another and with forces in the environment in ways difficult to detect and measure. The classic example in medicine is cancer. The classic example in ecology is the stabilization of ecosystems.

So scientists shuffle along as best they can, intuiting, guessing, tinkering, gaining more information along the way. They persist until solid explanations can be put together and a consensus emerges, sometimes quickly but at other times only after a long period.

When a phenomenon displays invariable properties under clearly defined conditions, then and only then can a scientific explanation be declared to be a scientific fact. The recognition that hydrogen is one of the elements, incapable of being divided into other substances, is a fact. That an excess of mercury in the diet causes one disease or another can, after enough clinical studies are conducted, be declared a fact. It may be widely believed that mercury causes an entire class of similar maladies, due to the one or two known chemical reactions in cells of the body. This idea may or may not be confirmed by further studies on diseases believed affected in this manner by mercury. Meanwhile, however, when research is still incomplete, the idea is a theory. If the theory is proved wrong, it was not necessarily also altogether a bad theory. At least it will have stimulated new research, which adds to knowledge. That is why many theories, even if they fail, are said to be "heuristic"—they are good for the promotion of discovery. Incidentally, the source of the word *eureka*—"I have found it!"—descends from the legend of the Greek scientist Archimedes, who, while sitting in a public bath, imagined how to measure the density of an object regardless of its shape. Put it in water, measure its volume by the rise in the water level, and its weight by how fast it sinks in the water. The density is the amount of weight divided by the amount of volume. Archimedes is said to have then left the bath, running through the streets, hopefully in his robe, while shouting, *Heurika!* Specifically, he'd found how to determine whether a crown was pure gold. The pure substance has a higher density than gold mixed with silver, the lesser of the two noble metals. But of far greater importance, Archimedes had discovered how to measure the density of all solids regardless of their shape or composition.

Now consider a much grander example of the

scientific method. It has been commonly said, all the way back to the publication of Charles Darwin's *On the Origin of Species* in 1859, that the evolution of living forms is just a theory, not a fact. What could have been said already from evidence in Darwin's time, however, was that evolution is a fact, that it has occurred in at least some kinds of organisms some of the time. Today the evidence for evolution has been so convincingly documented in so many kinds of plants, fungi, animals, and microorganisms, and in such a great array of their hereditary traits, coming from every discipline of biology, all interlocking in their explanations and with no exception yet discovered, that evolution can be called confidently a fact. In Darwin's time, the idea that the human species descended from early primate ancestors was a hypothesis. With massive fossil and genetic evidence behind it, that can now also be called a fact. What remains a theory still is that evolution occurs universally by natural selection, the differential survival and successful reproduction of some combinations of hereditary traits over others in breeding populations. This proposition has been tested so many times and in so many ways, it also is now close to deserved recognition as an established fact. Its implication has been and remains of enormous importance throughout biology.

When a well-defined and precisely consistent

process is observed, such as ions flowing in a magnetic field, a body moving in airless space, and the volume of a gas changing with temperature, the behavior can be precisely measured and mathematically defined as a law. Laws are more confidently sought in physics and chemistry, where they can be most easily extended and deepened by mathematical reasoning. Does biology also have laws?

I have been so bold in recent years as to suggest that, yes, biology is ruled by two laws. The first is that all entities and processes of life are obedient to the laws of physics and chemistry. Although biologists themselves seldom speak of the connection, at least in such a manner, those working at the level of the molecule and the cell assume it to be true. No scientist of my acquaintance believes it worthwhile to search for what used to be called the élan vital, a physical force or energy unique to living organisms.

The second law of biology, more tentative than the first, is that all evolution, beyond minor random perturbations due to high mutation rates and random fluctuations in the number of competing genes, is due to natural selection.

A source of the ground strength of science are the connections made not only variously *within* physics, chemistry, and biology, but also *among* these primary disciplines. A very large question remains in science and philosophy. It is as follows: Can this consilience—connections made between widely separated bodies of knowledge—be extended to the social sciences and humanities, including even the creative arts? I think it can, and further I believe that the attempt to make such linkages will be a key part of intellectual life in the remainder of the twenty-first century.

Why do I and others think in this controversial manner? Because science is the wellspring of modern civilization. It is not just "another way of knowing," to be equated with religion or transcendental meditation. It takes nothing away from the genius of the humanities, including the creative arts. Instead it offers ways to add to their content. The scientific method has been consistently better than religious beliefs in explaining the origin and meaning of humanity. The creation stories of organized religions, like science, propose to explain the origin of the world, the content of the celestial sphere, and even the nature of time and space. These mythic accounts, based mostly on the dreams and epiphanies of ancient prophets, vary from one religion's belief to another. Colorful they are, and comforting to the minds of believers, but each contradicts all the others. And when tested in the real world they have so far proved wrong, always wrong.

The failure of the creation stories is further evidence that the mysteries of the universe and the human mind cannot be solved by unaided intuition. The scientific method alone has liberated humanity from the narrow sensory world bequeathed it by our prehuman ancestors. Once upon a time humans believed that light allowed them to see everything. Now we know that the visual spectrum, which activates the visual cortex of the brain, is only a sliver of the electromagnetic spectrum, where the frequencies range across many orders of magnitude, from those of extreme high-frequency gamma rays at one end to those at the extreme low-frequency radiation at the other. The analysis of the electromagnetic spectrum has led to an understanding of the true nature of light. Knowledge of its totality has made possible countless advances in science and technology.

Once people thought that Earth was the center of the universe and lay flat and unmoving while the sun rotated around it. Now we know that the sun is a star, one of two hundred million in the Milky Way galaxy alone. Most hold planets in their gravitational thrall, and many of these almost certainly resemble Earth. Do the Earthlike planets also harbor life? Probably, in my opinion, and, thanks to the scientific method, furnished with improved optics and spectroscopic analyses, we will know in a short time.

Once it was believed that the human race arose full-blown in its present form as a supernatural event. Now we understand, in sharp contrast, that our species descended over six million years from African apes that were also the ancestors of modern chimpanzees.

As Freud once remarked, Copernicus demonstrated that Earth is not at the center of the universe, Darwin that we are not the center of life, and he, Freud, that we are not even in control of our own minds. Of course, the great psychoanalyst must share credit with Darwin, among others, but the point is correct that the conscious mind is only part of the thinking process.

Overall, through science we have begun to answer in a more consistent and convincing way two of the great and simple questions of religion and philosophy: Where do we come from? and, What are we? Of course, organized religion claims to have answered these questions long ago, using supernatural creation stories. You might then well ask, can a religious believer who accepts one such story still do good science? Of course he can. But he will be forced to split his worldview into two domains, one secular and the other supernatural, and stay within the secular domain as he works. It would not be difficult for him to find endeavors in scientific research that have no immediate relation to theology. This suggestion is not meant to be cynical, nor does it imply a closing of the scientific mind.

If proof were found of a supernatural entity or force that affects the real world, the claim all organized religions make, it would change everything. Science is not inherently against such a possibility. Researchers in fact have every reason to make such a discovery, if any such is feasible. The scientist who achieved it would be hailed as the Newton, Darwin, and Einstein, all put together, of a new era in history. In fact, countless reports have been made throughout the history of science that claim evidence of the supernatural. All, however, have been based on attempts to prove a negative proposition. It usually goes something like this: "We haven't been able to find an explanation for such-and-such a phenomenon; therefore it must have been created by God." Present-day versions still circulating include the argument that because science cannot yet provide a convincing account of the origin of the universe and of the setting of the universal physical constants, there must be a divine Creator. A second argument heard is that because some molecular structures and reactions in the cell seem too complex (to the author of the argument, at least) to have been assembled by natural selection, they must have been designed by a higher intelligence. And one more: because the human mind, and especially free will as a key part of it, appear beyond the capability of the material cause and effect, they must have been inserted by God.

The difficulty with reliance on negative hypotheses to support faith-based science is that if they are wrong, they are also very vulnerable to decisive disproof. Just one testable proof of a real, physical cause destroys the argument for a supernatural cause. And precisely this in fact has been a large part of the history of science, as it has unfolded, phenomenon by phenomenon. The world rotates around the sun, the sun is one star out of two hundred million or more in one galaxy out of hundreds of billions of galaxies, humanity descended from African apes, genes change by random mutations, the mind is a physical process in a physical organ. Yielding to naturalistic, real-world understanding, the divine hand has withdrawn bit by bit from almost all of space and time. The remaining opportunities to find evidence of the supernatural are closing fast.

As a scientist, keep your mind open to any possible phenomenon remaining in the great unknown. But never forget that your profession is exploration of the real world, with no preconceptions or idols of the mind accepted, and testable truth the only coin of the realm.

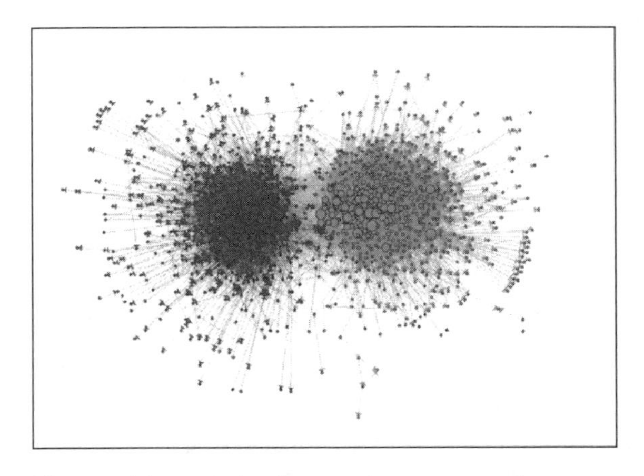

The potential community of contacts in contemporary human relationships (lines) is illustrated by political blogs (dots) in the 2004 U.S. presidential election. The same applies to disciplines of science. Modified from "The political blogosphere and the 2004 U.S. election: divided they blog," by Lada A. Adamic and Natalie Glance, *Proceedings of the 3rd International Workshop on Link Discovery (Link KDD'05)* 1: 36–43 (2005).

Five

The Creative Process

To know how scientists engage in visual imagery is to understand how they think creatively. Practicing it yourself while you receive your technical training will bring you close to the heart of the scientific enterprise. When earlier I said you can surely succeed, I also assumed that you are able to daydream. But be prepared mentally for some amount of chaos and failure. Waste and frustration often attend the earliest stages. When a workable idea emerges, the research becomes more routine, and also much easier to think about and explain to others. This is the part I have always enjoyed the most.

Since so much of good science—and perhaps all of great science—has its roots in fantasy, I suggest that you yourself engage in a bit right now. Where would you like to be, what would you most like to be doing professionally ten years from now, twenty years, fifty? Next, imagine that you are much older and looking back on a successful career. What kind of great discovery, and in what field of science, would you savor most having made?

I recommend creating scenarios that end with goals, then choosing ones you might wish to pursue. Make it a practice to indulge in fantasy about science. Make it more than just an occasional exercise. Daydream a lot. Make talking to yourself silently a relaxing pastime. Give lectures to yourself about important topics that you need to understand. Talk with others of like mind. By their dreams you shall know them.

Speaking of dreams, I once had dinner with Michael Crichton, the renowned thriller and science fiction writer. We talked about our respective professions. The movie *Rising Sun*, based on his book of the same name, had recently been released, and at the time we met it was stirring criticism over its perceived political message. The plot was about the effort of a Japanese high-tech corporation to expand its control in American industry by espionage and cover-up. At the time of the movie's release (1993), the Japanese economy was surging and its companies were buying pieces of America, from Rockefeller

Center to Hawaiian real estate. The overreaching theme that might be read into the story was that Japan, having failed to build an empire through force, was now trying to build one through economic dominance.

Crichton knew of earlier struggles over my 1975 book *Sociobiology: The New Synthesis*, which created a firestorm of protest from social scientists and radical leftist writers. They were incensed by my argument that human beings have instincts, and therefore that a gene-based human nature exists. At times the protest reached the level of interruption of my classes and public demonstrations. One in Harvard Square demanded my dismissal from Harvard.

Crichton asked, "How did you handle all that pressure?" It was embarrassing at times for me and my family, I said, but intellectually not difficult. It was obviously a contest of science against political ideology, and past history has shown that if the research is sound, science always eventually comes out on top. And it did this time, in favor of sociobiology, already at the time of our dinner conversation a well-established discipline. I suggested that the controversy over *Rising Sun*, which in any case is a work of fiction, was not a bad thing. It helped to sharpen different viewpoints over an important issue. Better to let it play out than encouraged to fester.

I took the opportunity to share with Crichton a thought experiment I had conducted that had been stimulated by his book and the movie *Jurassic Park*, the latter released the same year as *Rising Sun*. In *Jurassic Park* a billionaire hires a paleontologist and other experts to create dinosaurs for a park he wants to set up. This being science fiction, the project of course succeeds. The method devised was ingenious. First acquire pieces of amber formed as fossilized tree resin at the time of dinosaurs. Some of the fragments will contain well-preserved remains of mosquitoes. That much works in principle: I've studied hundreds of real fossil ants in amber from the Cretaceous Period, near the end of the Age of Dinosaurs. The next step in the plot was to find mosquitoes that still hold remnants of blood sucked from the veins of dinosaurs. Extract the dinosaur DNA they contain, and implant it in chicken eggs to grow dinosaurs. This is good science fiction. Each step verges on the far end of probability even though it is almost (notice that as a scientist I say almost!) certainly impossible.

I told Crichton of a somewhat similar experiment I had imagined that was really and truly possible. In the Harvard collection are large numbers of ants preserved in amber from the Dominican Republic, roughly twenty-five million years in age (younger than hundred-million-year-old dinosaurs, but still *old*). I had analyzed this fossil collection thoroughly and described a number of species new to science. Among these the most abundant was one I named *Azteca alpha*. A living species *Azteca muelleri*, which appears to be a direct evolutionary descendant or otherwise close relative of *Azteca alpha*, still lives in Central America. These ants use large quantities of pheromones, acrid-smelling terpenoids, which they release into the air to alarm nestmates whenever the colony is threatened by invaders.

I told Crichton that I might be able to extract remnants of the pheromone from the *Azteca alpha* remains, inject them into an *Azteca muelleri* nest, and get the alarm response. In other words, I could deliver a message from one ant colony to another across a span of twenty-five million years. This had Crichton's attention. He asked if I planned to do it. I said, not yet. I didn't have time, and still don't. In this particular dream there is too much of the circus trick and too little of real science—too little chance, that is, to discover something really new.

I'll end this letter by telling you how I

conceive of the creative process of both a novelist like Crichton and a scientist. (I have been both.) The ideal scientist thinks like a poet and only later works like a bookkeeper. Keep in mind that innovators in both literature and science are basically dreamers and storytellers. In the early stages of the creation of both literature and science, everything in the mind is a story. There is an imagined ending, and usually an imagined beginning, and a selection of bits and pieces that might fit in between. In works of literature and science alike, any part can be changed, causing a ripple among the other parts, some of which are discarded and new ones added. The surviving fragments are variously joined and separated, and moved about as the story forms. One scenario emerges, then another. The scenarios, whether literary or scientific in nature, compete with one another. Some overlap. Words and sentences (or equations or experiments) are tried to make sense of the whole thing. Early on, an end to all the imagining is conceived. It arrives at a wondrous denouement (or scientific breakthrough). But is it the best, is it true? To bring the end safely home is the goal of the creative mind. Whatever that might be, wherever located, however expressed, it begins as a phantom that rises, gains detail, then at the last moment either fades to be replaced, or, like the mythical giant Antaeus touching Mother Earth, gains strength. Inexpressible thoughts throughout flit along the edges. As the best fragments solidify, they are put in place and moved about, and the story grows until it reaches an inspired end.

A fire ant laying an odor trail. Drawing by Thomas Prentiss. Modified from "Pheromones," by Edward O. Wilson, *Scientific American* 208(5): 100–114 (May 1968).

Six

What It Takes

If you choose a career in science, and particularly in original research, nothing less than an enduring passion for your subject will last the remainder of your career, and life. Too many Ph.D.s are creatively stillborn, with their personal research ending more or less with their doctoral dissertations. It is you who aim to stay at the creative center whom I will now specifically address. You will commit your career, some good part of it, to being an explorer. Each advance in research you achieve will be measured, as scientists constantly do among themselves, by completing one or more of the following sentences:

"He [or she] discovered that . . ."

"He [or she] helped to develop the successful theory of . . ."

"He [or she] created the synthesis that first tied together the disciplines of . . ."

Original discoveries cannot be made casually, not by anyone at any time or anywhere. The frontier of scientific knowledge, often referred to as the cutting edge, is reached with maps drawn by earlier investigators. As Louis Pasteur said in 1854, "Fortune favors only the prepared mind." Since he wrote this, the roads to the frontier have greatly lengthened, and there is an enormously larger population of scientists who travel to get there. There is a compensation for you in your journey, however. The frontier is also vastly wider now, and it grows more so constantly. Long stretches along it remain sparsely populated, in every discipline, from physics to anthropology, and somewhere in these vast unexplored regions you should settle.

But, you may well ask, isn't the cutting edge a place only for geniuses? No, fortunately. Work accomplished on the frontier defines genius, not just getting there. In fact, both accomplishments along the frontier and the final eureka moment are achieved more by entrepreneurship and hard work than by native intelligence. This is so much the case that in most fields most of the time, extreme brightness may be a detriment. It has occurred to me, after meeting so many successful researchers in so many disciplines, that the ideal scientist is smart only to an intermediate degree: bright enough to see what can be done but not so bright as to become bored doing it. Two of the most original and influential Nobel Prize winners for whom I have such information, one a molecular biologist and the other a theoretical physicist, scored IQs in the low 120s at the start of their careers. (I personally made do with an underwhelming 123.) Darwin is thought to have had an IQ of about 130.

What, then, of certified geniuses whose IQs exceed 140, and are as high as 180 or more? Aren't they the ones who produce the new groundbreaking ideas? I'm sure some do very well in science, but let me suggest that perhaps, instead, many of the IQ-brightest join societies like MENSA and work as auditors and tax consultants. Why should the rule of optimum medium brightness hold? (And I admit this perception of mine is only speculative.) One reason could be that IQ-geniuses have it too easy in their early training. They don't have to sweat the science courses they take in college. They find little reward in the necessarily tedious chores of data-gathering and analysis. They choose not to take the hard roads to the frontier, over which the rest of us, the lesser intellectual toilers, must travel.

Being bright, then, is just not enough for those who dream of success in scientific research.

Mathematical fluency is not enough. To reach and stay at the frontier, a strong work ethic is absolutely essential. There must be an ability to pass long hours in study and research with pleasure even though some of the effort will inevitably lead to dead ends. Such is the price of admission to the first rank of research scientists.

They are like treasure hunters of older times in an uncharted land, these elite men and women. If you choose to join them, the adventure is the quest, and discoveries are your silver and gold. How long should you keep at it? As long as it gives you personal fulfillment. In time you will acquire world-class expertise and with certainty make discoveries. Maybe big ones. If you are at all like me (and almost all the scientists I know are, in this regard), you will find friends among your fellow enthusiasts and experts. Daily satisfaction from what you are doing will be one of your rewards, but of equal importance is the esteem of people you respect. Yet another is the recognition that what you find will uniquely benefit humanity. That alone is enough to kindle creativity, though it cannot alone sustain it.

How hard will this be? I'll pull no punches about that part. At Harvard I advised mostly graduate students who planned for academic careers. They chose to combine research with teaching in a research university or liberal arts college. I posited the following time for success in this combination: at the start, forty hours a week for teaching and administration; up to ten hours for continued study in your specialty and related fields; and at least ten hours in research—presumably in the same field as your Ph.D. or postdoctoral work, or close enough to draw on the experience from your student years. Sixty hours a week total can be daunting, I know. So seize every opportunity to take sabbaticals and other paid leaves that allow you stretches of full-time research. Avoid department-level administration beyond thesis committee chairmanships if at all fair and possible. Make excuses, dodge, plead, trade. Spend extra time with students who show talent and interest in your field of research, then employ them as assistants for your benefit and theirs. Take weekends off for rest and diversion, but no vacations. Real scientists do not take vacations. They take field trips or temporary research fellowships in other institutions. Consider carefully job offers from other universities or research institutions that include more research time and fewer teaching and administrative responsibilities.

Don't feel guilty about following this advice. University faculties consist of both "inside professors," who enjoy work that involves close social interactions with other faculty members and take justifiable pride in their service to the institution, and "outside professors," whose social interactions are primarily with fellow researchers. Outside professors are light on committee work but earn their keep another way: they bring in a flow of new ideas and talent and they add prestige and income proportionate to the amount and quality of their discoveries.

Wherever your research career takes you, whether into academia or otherwise, stay restless. If you are in an institution that encourages original research and rewards you for it, stay there. But continue to move about intellectually in search of new problems and new opportunities. Granted that happiness awaits those who can find pleasure while working on the same subject all their careers, and they assuredly have a good chance of making breakthrough advances while doing so. Polymer chemistry, computer programs of biological processes, butterflies of the Amazon, galactic maps, and Neolithic sites in Turkey are the kinds of subjects worthy of a lifetime of devotion. Once deeply engaged, a steady stream of small discoveries is guaranteed. But stay alert for the main chance that lies to the side. There will always be the possibility of a major strike, some wholly unexpected find, some little detail that catches your peripheral attention that might very well, if followed, enlarge or even transform the subject you have chosen. If you sense such a possibility, seize it. In science, gold fever is a good thing.

To make such success more likely, there is another

quality in which you might or might not be well endowed but if not should at least try to cultivate. It is entrepreneurship, the willingness to try something daunting you've imagined doing and no one else has thought or dared. It could be, for example, starting a project in a part of the world neither you nor your colleagues have yet visited; or finding a way to try an already available instrument or technique not yet used in your field; or, even more bravely, applying your knowledge to another discipline not yet exposed to it.

Entrepreneurship is enhanced by performing lots of quick, easily performed experiments. Yes, that's what I just said: experiments quick and easily performed. I know that the popular image of science is one of uncompromising precision, with each step carefully recorded in a notebook, along with periodic statistical tests on data made at regular intervals. Such is indeed absolutely necessary when the experiment is very expensive or time-consuming. It is equally demanded when a preliminary result is to be replicated and confirmed by you and others in order to bring a study to conclusion. But otherwise it is certainly all right and potentially very productive just to mess around. Quick uncontrolled experiments are very productive. They are performed just to see if you can make something interesting happen. Disturb Nature and see if she reveals a secret. To show you my own devotion to the quick and sloppy, I'll give you several examples from my own initially crude efforts. These are from memory; I didn't keep notes, careful or otherwise.

- I put a powerful magnet over a column of running ants to see if I could turn their direction or at least disrupt them, and hence detect whether ants have a magnetic sense. Time consumed: two hours. Result: failed. The ants couldn't care less.

- I sealed off the metapleural glands of ants in a laboratory colony. These tiny organs are clusters of cells found on each side of the middle part of the body. I then let the operated ants run over the screened roof of a culture of soil bacteria, and also over other cultures with ants not so treated, in order to see if the metapleural glands shed airborne antibiotic substances. Time consumed: two weeks. Result: failed. (I should have continued the effort, becoming more persistent and using different methods. The substances are there, as subsequent researchers showed.)

- I tried to create mixed colonies of two species of fire ants by chilling them and switching their queens. Time consumed: two hours. Result: success! I used the method to prove (with careful experiments and neat notes this time) that the traits distinguishing the two species are due to different genes. Chilling and mixing is now a standard technique for several lines of research.

- In the 1950s, it was thought that ants probably communicate with chemical signals (later called pheromones). But the possibility was still open that they use instead coded tappings and strokings with their antennae. A drumbeat of antennae on the body of a nestmate, for example, might be an alarm signal. I decided to see if I could locate the gland that produces odor trails. If successful, I thought, that could be the first step in working out the ant pheromone code. I dissected out all the main organs in the abdomen of worker fire ants and laid artificial trails made from them, patiently slicing and picking under the microscope with the finest surgical forceps. Time consumed: one week. Result: there was no response to any of the first organs tried, but then to my surprise came a powerful response to the Dufour's gland, an almost invisible finger-shaped organ located at the base of the sting. A major success this time. Not only did the fire ants follow the trail, they rushed out of the nest to get onto and follow it. The Dufour's secretions, it seemed, are both guides and stimulants: this was a new concept in pheromone studies. Other scientists and I went on during the following years to work out the dozen or so pheromone signals that compose most of the ant vocabulary.

Performing small, informal experiments is an exciting sport, and the risk in lost time is small. However, if a preliminary procedure proves necessarily time-consuming or expensive or both, the cost in time and money can become quickly prohibitive. If the effort fails, entrepreneurship requires the character and the means to start over—just as it does in business and other careers outside of science.

I will close this letter with one further piece of relevant practical advice to offer you if you are already a graduate student or young professional. Unless your training and research commit you to a major research facility, for example a supercollider, space telescope, or stem-cell laboratory, do not linger too long with any one technology. When a new instrument is at the cutting edge, it may open new horizons of research quickly, but it is also at first usually expensive and difficult to operate. As a result, there will be a temptation for a young scientist to build a career in the new technology itself rather than to make original studies that can be performed with it. In biochemistry and cell biology, for example, the centrifuge has long been essential for spinning apart different kinds of molecules and by this means making them available for physical and chemical analysis. In this way the trees can be separated from the forest, so to speak, and by this means can make the whole forest more understandable. At the beginning, centrifuges required a room of their own and a trained technician to manage them. As their engineering was streamlined, however, any researcher could, with a few instructions, run the machines alone. Then centrifuges came out of their personal laboratories in the form of smaller, less expensive units. Today, graduate students in many fields of biology accept them as a routine part of their tabletop armamentarium. The same progression, from technology worthy of a discipline of its own to a routine part of every well-equipped laboratory, also occurred in the evolution of scanning electron microscopy, electrophoresis, computers, DNA sequencing, and inferential statistics software.

The principle I have drawn from this history is the following: *use but don't love technology*. If you need it but find it at all forbiddingly difficult, recruit a better-prepared collaborator. Put the project first and, by any available and honorable means, complete and publish the results.

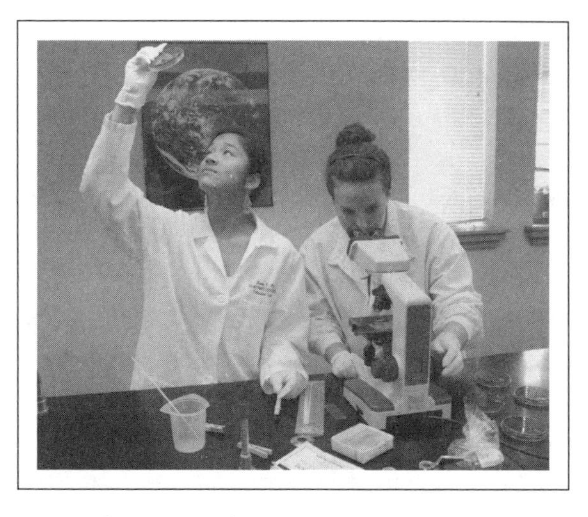

At the Alabama School of Mathematics and Science (ASMS), Allison Kam (left) and Hannah Waggerman examine environmental bacteria samples taken from the Mobile Delta. Photograph by John Hoyle.

Seven

Most Likely to Succeed

How are born scientists best discovered? There is a growing movement to identify secondary school students of promise and open to them special curricula that encourage talent. One example I know about personally is the Alabama School of Mathematics and Science in my home town of Mobile, which selects high school students from all over the state, provides them with scholarships, and settles them in a resident college-like campus. Immersed in laboratory research guided by experienced scientists, students learn in an atmosphere where a focus on science and technology is the norm. Virtually all of the graduates in a given year thus far have gone straight to college.

Few scientists write memoirs, and among those who do, even fewer are willing to disclose the emotions, urges, idols, and teachers that brought them into their scientific careers. In any case, I don't trust most such accounts, not because the authors are dishonest, but because the scientific culture discourages such disclosures. Scientific researchers have a hard enough time avoiding any utterance that might sound childish, poetic, or dilatory and insubstantial to other scientists. Hence a leathery, just-the-facts style confines most personal accounts of scientific discovery, and a good story often comes out reticent and dull. False modesty is the peccadillo of the scientific memoirist.

An example (imaginary) might read as follows: "While working at the Whitehead Institute X-ray crystallography laboratory on avian muscle protein, I became fascinated with the classical problem of autonomous folding. I was led to consider . . ."

Well, I'm sure that such writers in real life were fascinated and even compelled to consider this or that, but not me reading their account. A reader would like to know the reason why they did the hard work to achieve their goal. Where was the adventure, what was the dream?

So there is a great deal we don't know about what makes scientists, and how they really feel about their work. Without the Alabama School of Mathematics and Science, would the elite students there all have gone to college and careers related to science?

Another question is whether it is more inspiring and useful for such students to work in small teams or on individual projects that each selects, however idiosyncratic. We have no clear answer to either of these questions. But I have no doubt that encouragement given teenagers who are already predisposed to scientific careers does help lead them to success in later years.

Basically this question about teams arises in the encouragement of innovation by practicing scientists. The conventional wisdom holds that science of the future will be more and more the product of "teamthink," multiple minds put in close contact. It is certainly the case that fewer and fewer solitary authors publish research articles in premier journals such as *Nature* and *Science*. The number of coauthors is more often three or more; and in the case of a few subjects, such as experimental physics and genome analysis, where research by necessity involves an entire institution, the number sometimes soars to over a hundred.

Then there are the vaunted science and technology think tanks, where some of the best and brightest are brought together explicitly to create new ideas and products. I've visited the Santa Fe Institute in New Mexico, as well as the development divisions of Apple and Google, two of America's corporate giants, and I admit I was very impressed with their futuristic ambience. At Google I even commented, "This is the university of the future."

The idea in these places is to feed and house very smart people and let them wander about, meet in small groups over coffee and croissants, and bounce ideas off each other. And then, perhaps while strolling through well-manicured grounds or on their way to a gourmet lunch, they will experience the flash of epiphany. This surely works, especially if there is a problem in theoretical science already well formulated, or else a product in need of being designed.

But is groupthink the best way to create really new science? Risking heresy, I hereby dissent. I believe the creative process usually unfolds in a very different way. It arises and for a while germinates in a solitary brain. It commences as an idea and, equally important, the ambition of a single person who is prepared and strongly motivated to make discoveries in one domain of science or another. The successful innovator is favored by a fortunate combination of talent and circumstance, and is socially conditioned by family, friends, teachers, and mentors, and by stories of great scientists and their discoveries. He (or she) is sometimes driven, I will dare to suggest, by a passive-aggressive nature, and sometimes an anger against some part of society or problem in the world. There is also an introversion in the innovator that keeps him from team sports and social events. He dislikes authority, or at least being told what to do. He is not a leader in high school or college, nor is he likely to be pledged by social clubs. From an early age he is a dreamer, not a doer. His attention wanders easily. He likes to probe, to collect, to tinker. He is prone to fantasize. He is not inclined to focus. He will not be voted by his classmates most likely to succeed.

When prepared by education to conduct research, the most innovative scientists of my experience do so eagerly and with no prompting. They prefer to take first steps alone. They seek a problem to be solved, an important phenomenon previously overlooked, a cause-and-effect connection never imagined. An opportunity to be the first is their smell of blood.

On the frontier of modern science, however, multiple skills are almost always needed to bring any new idea to fruition. An innovator may add a mathematician or statistician, a computer expert, a natural-products chemist, one or several laboratory or field assistants, a colleague or two in the same specialty—whoever it takes for the project to succeed becomes a collaborator. The collaborator is often another innovator who has been toying with the same idea, and is prone to modify or add to it. A critical mass is achieved and discussion intensifies, perhaps among scientists in the same place, perhaps scattered around the world. The project moves forward until an original result is achieved. Group thought has brought it to fruition.

Innovator, creative collaborator, or facilitator: in the course of your successful career, you may well fill each of these roles at one time or another.

The author with sweep net looking at insects: Mobile, Alabama, 1942 (left), and the summit of Gorongosa Mountain, Mozambique, 2012 (right). Photographers: 1942, Ellis MacLeod; 2012, © Piotr Naskrecki.

Eight

I Never Changed

APPROACHING THE END of more than sixty years of research, I have been fortunate to have been given complete freedom in choosing my subjects. Because I no longer look to very much in the way of a future, and the fires of decent ambition have been accordingly damped, I can tell you, without the debilitating drag of false modesty, how and why some of my discoveries were made. I'd like you to think, as I thought early in my career of older scientists, "If he could do it, so can I, and maybe better."

I started very young, even before my snake-handling triumph in Camp Pushmataha. Maybe you started young too, or else you are young and just starting. Back in 1938 when I was nine years old, my family moved from the Deep South to Washington, D.C. My father was called there for a two-year stint as an auditor in the Rural Electrification Administration, a Depression-era federal agency charged with bringing electric power to the rural South. I was an only child, but not especially lonely. Any kid that age can find a buddy or fit into some small neighborhood group, maybe at the risk of a fistfight with the alpha boy. (For years I carried scars on my upper lip and left brow.) Nevertheless, I was alone that first summer and was left to my own devices. No stifling piano lessons, no boring visits to relatives, no summer school, no guided tours, no television, no boys' clubs, nothing. It was *wonderful*! I was enchanted at this time by Frank Buck movies I'd seen about his expeditions to distant jungles to capture wild animals. I also read *National Geographic* articles that told about the world of insects—big metallic-colored beetles and garish butterflies, also mostly from the tropics. I found an especially absorbing piece in a 1934 issue entitled "Ants, Savage and Civilized," which led me to search for these insects—searches that were always successful due to the overwhelming abundance of ants everywhere I looked.

There were postage stamps to collect and comic books, of course, but also butterflies and ants. Nothing complicated about collecting and studying insects. For a while anyway, they served as my lions and tigers, not exactly big game snared in nets by a hundred native assistants, but nevertheless the real thing. Thus fired up, I put some bottles in a cloth bag and walked over to the nearby woods of Rock Creek Park on my first expedition, venturing into second-growth deciduous woodland crisscrossed by paths. I remember vividly the animals I brought home that day. They included a wolf spider and the red and green nymph of a long-horned grasshopper.

Subsequently I decided to add butterflies as my quarry. My stepmother made me a butterfly net. (I put together a lot of them in the years to follow. In case you would like to do the same, bend a wire coat hanger into a circular loop, straighten the hook, heat the hook until it can burn wood, then push it into the end of a sawed-off broomstick. Finally, sew a net of cheesecloth or mosquito netting around the hoop.)

Thus accoutered, my butterfly collection grew furiously. Early in this career of mine, my best friend Ellis MacLeod, who years later was to be a professor of entomology at the University of Illinois, told me he had seen a medium-sized butterfly, black with brilliant red stripes across both wings, fluttering back and forth around the bushes in front of his apartment building. We found a book on butterflies and identified it as Red Admiral. The book was the beginning of my library on insects. At this point my mother, living with her second husband in Louisville, Kentucky, sent me a larger, beautifully illustrated book on butterflies. It threw me into confusion. The only familiar species I found in it was the Cabbage White, a species accidentally introduced from Europe many years previously. The reason for my confusion, I learned later, was the book was about British butterflies.

My future was set. Ellis and I agreed we were going to be entomologists when we grew up. We delved into college-level textbooks, which we could scarcely read, although we tried very hard. One that we checked out from a public library and worked on page by page was Robert E. Snodgrass's formidable *Principles of Insect Morphology*, published in 1935. Only later did I learn that grown-up biologists were using it as a technical reference book. We visited the insect collections on display at the awesome National Museum of Natural History, aware that professional entomologists were curators there. I never saw one of these demigods (one was Snodgrass himself), but just knowing they were there as part of the *United States government* gave me hope that one day I might ascend to this unimaginably high level.

Returning in 1940 with my family to Mobile, I plunged into the rich new fauna of butterflies. The semitropical climate and nearby swamps were a close realization of my earlier dreams. To the red admirals, painted ladies, great spangled fritillaries, and mourning cloaks characteristic of the more northern climes I added snout butterflies, Gulf fritillaries, Brazilian skippers, great purple hairstreaks, and several magnificent swallowtails—giant, zebra, and spicebush.

Then I turned to ants, monomaniacally determined to find every kind living in the weed-grown vacant lot next to our large family house on Charleston Street. I didn't know the scientific names of the species, but I do now, and the location of every colony in the quarter-acre space is vivid in my memory: the Argentine ant (*Linepithema humile*), which nested in the rotting wooden fence at the edge of the lot in the winter and spread out among the weeds during the warm months; large black ants (*Odontomachus brunneus*) with snapping jaws and vicious stings, which inhabited a pile of roof shingles at the far corner beneath a fig tree; a huge mound-dwelling colony of the red imported fire ant (*Solenopsis invicta*) that I found at the edge of the lot next to the street; and a colony of a tiny yellow species (*Pheidole floridana*) nesting beneath an old whiskey bottle.

Three years later, as nature counselor at Pushmataha, I transitioned into a snake period, and began catching as many as I could find of the dozens of species that inhabit southwestern Alabama.

I've gone into this boyhood story to make a point that may be relevant to your own career trajectory. *I have never changed.*

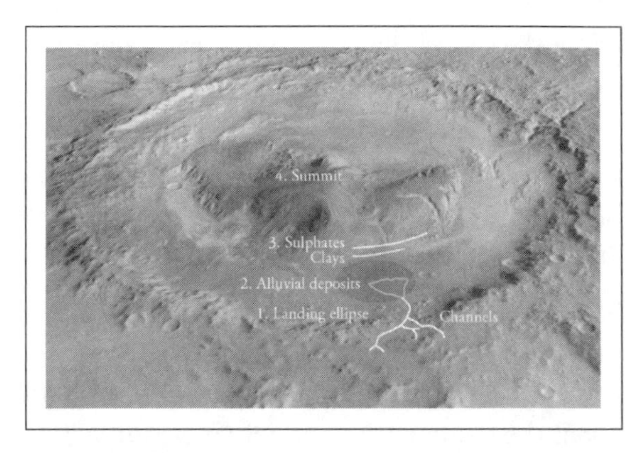

Planned path of the Mars rover *Curiosity* in Gale Crater. "NASA picks Mars landing site," by Eric Hand, *Nature* 475: 433 (July 28, 2011). Modified from photograph by NASA/JPL-CALTECH/ASU/UA.

Nine

Archetypes of the Scientific Mind

The better emotions of our nature are felt and examined and understood more deeply during maturity, but they are born and rage in full intensity during childhood and adolescence. Thereafter they endure through the rest of life, serving as the wellsprings of creative work.

I told you earlier that during the earliest steps to discovery the ideal scientist thinks like a poet. Only later does he work at the bookkeeping expected of his profession. I spoke of passion and decent ambition as forces that drive us to creative work. The love of a subject, and I say it again for emphasis, is meritorious in itself. By pleasure drawn from discovery of new truths, the scientist is part poet, and by pleasure drawn from new ways to express old truths, the poet is part scientist. In this sense science and the creative arts are foundationally the same.

I could say more to you about the metaphorical temple of science, could speak of its infinite chambers and galleries, could offer you additional instructions on how to find your way. But you will learn all that on your own as you progress. Better at this point to explore with you some of the psychology of innovation. I suggest that you examine your inner thoughts in broader terms to locate the kinds of satisfaction you might obtain from a career in science. The value of this exercise in self-analysis applies equally well to professions in research, teaching, business, government, and the media.

Psychologists have identified five components in personality, partly based on differences in genes, on which the inner lives of people are based. My impression is that research scientists are more prone to introversion as opposed to extroversion, are neutral (can go either way) to antagonism versus agreeableness, and lean strongly toward conscientiousness and openness to experience. The circumstances in their lives that bend them toward creative work vary enormously, and the events that spark their interest in particular research opportunities differ by at least as much.

Nevertheless, I will repeat my conviction that you will become most devoted to research in science and technology through images and stories that have affected you early—particularly from childhood to the fringes of post-adolescence, say from nine or ten years of age through the teenage years into the early twenties. Further, the transformative events can be classified into a relatively small number of general images that carry maximum long-term impact. I will call them archetypes, believing they are comparable to the imprinting that makes it easier to learn languages and mathematics at a relatively early age. Archetypes, as scholars have noted, are commonly expressed by stories in myth and the creative arts. They are also powerfully manifested in the great technoscientific enterprise. It will make a difference in your own creative life if you are moved by one or more.

THE JOURNEY TO AN UNEXPLORED LAND. This yearning takes a variety of forms: to search for an unknown island; to climb a distant mountain and look beyond; to journey up an unexplored river; to contact a tribe rumored to live there; to discover lost worlds; to find Shangri-la; to land on another planet; to settle and start life anew in a distant country.

In science and technology, this archetype is expressed variously in the urge to find new species in unexplored ecosystems; to map the microscopic structure of the cell; to locate unsuspected pheromones and hormones that link organisms and tissues together; to view the deepest part of Earth's seafloor; to travel along and map the contours of the tectonic plates and canyons; to peer on through inner Earth to the core; to see the outer boundary of the universe; to discover signs of life on other planets; to listen for alien messages on the SETI telescopes; to find ancient organisms in fossils that date back to the beginning of life on Earth; and to uncover the remains of our prehuman ancestors and thereby disclose at long last where we came from and what we are.

SEARCH FOR THE GRAIL. The grail exists in many forms: the powerful formula (or talisman) known to the ancients but lost or kept secret; the Golden Fleece; the symbol of the secret society; the philosopher's stone; the path to the center of Earth; the incantation that releases evil spirits; the formula for enlightenment of mind and transcendence of soul; the hidden treasure; the key that unlocks the otherwise unassailable gate; the fountain of youth; the rite or magical potion that confers immortality.

Proceeding to the real world and the goals of science, we find equivalents that rouse the spirit in a similar manner. The grail is the discovery of a new and powerful enzyme or hormone; breaking the genetic code; discovering the secret of the origin of life; finding evidence of the first organism that evolved; the creation of a simple organism in the laboratory; the attainment of human immortality; achieving controlled fusion power; solving the mystery of dark matter; detecting neutrinos and the Higgs boson; deducing wormholes and multiverses.

GOOD AGAINST EVIL. Our stronger myths and emotions are driven by war against alien invaders; the conquest of new lands by our own people (who of course we regard as the civilized, the virtuous, the godly, and the chosen against the savages opposing us); the war of God against Satan; the overthrow of an evil tyrant; the triumph of the Revolution against all odds; the Hero, the Champion, or the Martyr vindicated in the end; the inner struggle of conscience between right and wrong; the Good Wizard; the Good Angel; the Magical Force; arrest and punishment of the criminal; vindication of the whistle-blower.

In the real world of science, we are aroused by what we call the war against cancer; the fight against other deadly diseases; the conquest of hunger; the mastery of a new energy source that can save the world; the campaign against global warming; forensic DNA sequencing to capture a criminal.

These several archetypes resonate up from the deep roots of human nature. They are appealing and easily understood. They convey meaning and power to humanity's creation myths. They are retold in the epic stories of history. They are the themes of great dramas and novels.

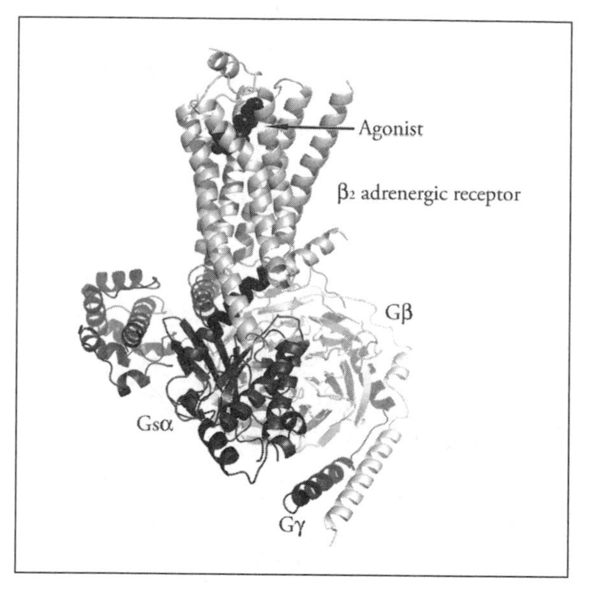

Cell-surface receptor activated by a signaling molecule (agonist, top) turns on a G-protein-coupled receptor that activates the G protein (3 G's, lower half). © Brian Kobilka.

Ten

Scientists as Explorers of the Universe

The Explorers Club of New York was founded in 1904 to celebrate the geographical exploration of the world and (later) outer space. Over the years the roster has included Robert Peary, Roald Amundsen, Theodore Roosevelt, Ernest Shackleton, Richard Byrd, Charles Lindbergh, Edmund Hillary, John Glenn, Buzz Aldrin, and other famous adventurers of the twentieth century. The headquarters of the Explorers Club on East Seventieth Street are stuffed with archives and memorabilia of the world's great wanderers. Also kept there are the famous expedition flags, carried over decades by members who journey to distant and sometimes virtually inaccessible destinations. When the explorer returns, so does the flag, along with an account of what was discovered.

Each year an annual dinner is held by the club at the Waldorf Astoria, a grand edifice evoking an era of great wealth. Dress is formal, and attendees are urged to wear whatever medals they have received in past exploits. It is the only occasion of which I am aware in North America where the latter embellishment is practiced. At dinner the excess of display turns to merriment. For years, until a guest became ill at one of the dinners, the fare was a humorous sample of what the explorer might be forced to eat when supplies run out: candied spiders, fried ants, crispy scorpions, broiled grasshoppers, roasted mealworms, exotic fish, and wild game.

In 2004 I was elected an honorary member, a distinction given only a score of men and women, and in 2009 I received the Explorers Club Medal, the highest award. At first this might be seen as an entirely inappropriate honor, and maybe it was. I had never suffered privation on polar ice, never climbed an unconquered Antarctic mountain, never contacted a previously unknown Amazonian tribe. The reason was science. The board of the Explorers Club had decided to expand its concept of what remains left to explore on our planet. The conventional map of the world had been largely filled in since the time Teddy Roosevelt traveled down an unnamed river in the Amazon and Robert Peary and Matthew Henson conquered the North Pole. Most of Earth's land surface had been visited on foot or by helicopter. What remained could be examined—even monitored day by day—through satellites to the last square kilometer. What was left of importance to map on the home planet other than the deep sea? The answer is its little-known biodiversity, that variety of plants, animals, and microorganisms that compose the thin layer of Earth called the biosphere. Although most of the flowering plants, birds, and mammals have been found, described, and given a scientific name, the great majority of species in other groups of organisms still remained to be discovered. Biologists and naturalists, both professional and amateur, who set out to find species and map the biosphere, have remained as among Earth's true explorers.

At the dinner in 2009 on which biodiversity was officially added to the worthy unknown, I had the extraordinary experience of giving the main address. There was much to be excited about that evening, but the memory that first comes to my mind was meeting the son of Tenzing Norgay, who in 1951, with Edmund Hillary, first summited Mount Everest. I reminded him that upon his return from the mountain, when a journalist had asked Tenzing Norgay, "How does it feel to be a great man?" he responded, "It is Everest that makes men great." To which I may add, to young biologists in particular who dream of combining science with physical adventure, it is the biosphere that offers you opportunities of epic proportion.

On Monday, July 3, 2006, the Explorers Club conducted its first "expedition" to explore biodiversity. It joined the American Museum of Natural History and other local nature-oriented organizations to conduct a bioblitz in New York City's Central Park. Bioblitzes are events in which experts on every kind of organism, from bacteria to birds, gather to find and identify as many species as possible during a stated short period of time, usually twenty-four hours. The aim on that day was to introduce the public to the concept that even a much-tramped-over urban area teems with the diversity of life. At the end of the day, the 350 registered volunteers had tallied—and mind you, this was in New York City—836 species, including 393 plants and 101 animals, the latter including 78 moths, 9 dragonflies, 7 mammals, 3 turtles, 2 frogs, and 2 microscopic, caterpillar-like tardigrades, the last enigmatic and seldom studied anywhere in the world. The tardigrades were the first ever reported from Central Park. One of the frogs was later determined to be a species new to science and found only in and around New York City.

On Tuesday, July 8, 2003, for the first time during any bioblitz, samples of soil and water were collected for later analysis of bacteria and other microorganisms, the most abundant and diverse of all forms of life. There was even physical adventure of a sort. Sylvia Earle, a leading marine biologist renowned for her dives in oceans around the world, offered to explore the murky slime-filled waters of the small lake next to the Bethesda Fountain, in order to add aquatic creatures to our list. "While I have had no concern," she observed, "about diving with sharks and killer whales or other creatures in the ocean, I did have reason to be mighty fearful of the microbes in the green pond in Central Park." She and others brave enough to dive with her produced a substantial list of species. There was one uncertain identification. "I found a snail floating by," Earle reported. "But I'm not sure if it was a resident or if it was introduced by the nearby restaurant as an escargot."

Very few places remain on Earth that are *not* seething with species of plants, animals, or microorganisms. At this time, for all intents and purposes the biological diversity seems almost infinite; and each living species in turn offers scientists boundless opportunities for important original research.

Consider a rotting tree stump in a forest. You and I casually walking past it on a trail would not give it more than a passing glance. But wait a moment. Walk around the stump slowly, look at it closely—as a fellow scientist. Before you, in miniature, is the equivalent of an unexplored planet. What you can learn from the decaying mass depends on your training and the science you have chosen to begin your career. Choose a subject, draw on it from anywhere in physics, chemistry, or biology. With imagination you will conceive original research programs that can be centered on the rotting stump.

Let's think about this more together. By research specialization I am a student of ecology and biodiversity. So join me in those overlapping domains of science, and let's ask: What life exists in the stump microplanet?

Start with animals. There may be cavities in the side, or at the base or beneath the roots, large enough to hold a mouse-sized mammal, and if not, surely a frog, salamander, snake, or lizard. Let us next magnify the image to bring in insects and other invertebrates one millimeter to thirty millimeters in length. We can see most of them with unaided vision. They are each distributed according to niches for which millions of years of evolution have adapted them. A large minority are insects. An entomologist trained in taxonomy (as should also be the case for any other scientist who needs to tell one species from another) will point out beetles that live here—members of the taxonomic families Carabidae (common name: ground beetles), Scarabaeidae (scarabs), Tenebrionidae (darkling beetles), Curculionidae (weevils),

Scydmaenidae (antlike stone beetles), and several others. More species of beetles are known than any other comparable group of organisms in the world. Yet even though the most diverse, they are not the most abundant in individuals. If the stump is well along in decomposition, ant colonies will be there, resting in the frass beneath the bark and among the roots below. Termites may riddle the heartwood. In the crevice and over the surface can be found bark lice, springtails, proturans, fly and moth larvae, earwigs, japygids, and symphylans. Around them a myriad of other rotting-stump invertebrates other than insects: crustacean pill bugs, tiny annelid worms, centipedes of varying sizes and shapes, slugs, snails, pauropods, and a huge fauna of mites, the numbers of the latter dominated by sluggish spherical oribatids with a sprinkling of wolfish, fast-running phytosciids. Spiders of many kinds spin webs or hunt widely on foot.

In patches of moss and lichens that grow on the surface of the stump—little worlds of their own—roam the aforementioned tardigrades, also called bear-animalcules for their body shape midway between caterpillars and miniature bears. Among these animals are the most abundant of all: the nematodes, also called roundworms, most barely visible. Worldwide, roundworms are reckoned to make up four-fifths of all the individual animals.

If my staccato listing confuses you, like a page

torn from a telephone book, rest assured it also confuses most biologists as well, and yet it is only the beginning of a very long roster that could be called out from our stump.

Throughout the decaying wood, fungal strands penetrate, the hyphae hanging in gossamer strands when the bark is pulled free. Microscopic fungi abound wherever there is moisture. Ciliates and other protistans swim in films and droplets of water.

All of the life of the stump ecosystem is dwarfed, however, in both variety and numbers of organisms, by the bacteria. In a gram of detritus on the surface or soil beneath the stump's base exist a billion bacteria. Together this multitude represents an estimated five thousand to six thousand species, virtually all unknown to science. Still smaller and likely even more diverse and abundant (we don't know for sure) are the viruses. To give you a sense of relative size at this lowest end of the stump-world scale, think of one cell of a multicellular organism as the size of a small city. A bacterium would then be the size of a football field and a virus the size of a football.

Yet—all of this ensemble, as we pause next to it for an hour or a day, is no more than a snapshot. Across a period of months and years, as the stump decays further, there is a gradual change of species, the numbers of organisms in each species, and the niches they fill. During the transition, new niches open and old ones close as the stump evolves from hard fresh-cut wood leaking resin to rotting splinters leaking nutrients into the soil. Finally, the stump becomes no more than crumbled fragments and mold, infiltrated by roots of invading neighbor plants and covered by dead twigs and leaf litter fallen from the canopy of the trees above. Throughout, the stump is a miniature ecosystem.

At each stage of decomposition, the stump's fauna and flora have been changing. In each cubic centimeter of its living and inert mass, the system has been passing energy and organic matter back and forth to the surrounding environment.

What could you make of this special world, should you choose to become an ecologist or biodiversity scientist and study it? How would you and your fellow researchers encompass the nearly infinite variations in Earth's biosphere represented by this microcosm? So much has been written, yet so very little is known—even the full census of stump-dwelling species and those of countless other kinds of miniature ecosystems on the land and in the sea remain unknown, unrecorded, unwritten. Drastically less has been learned of the lives and roles of each of the species in turn. Their combined order and process exceeds everything of which we have knowledge in the rest of the universe.

Keep in mind that a distinguished career of scientific research can be built from any one of the species, by means of contributions to different disciplines within biology, chemistry, and even physics. Karl von Frisch, the great German entomologist who made many discoveries concerning the honeybee, including their symbolic waggle-dance communication and their remarkable memory of place, knew that he had only begun to explore the biology of this single insect species. "The honeybee is like a magic well," he said. "The more you draw, the more there is to draw."

III

A LIFE *in* SCIENCE

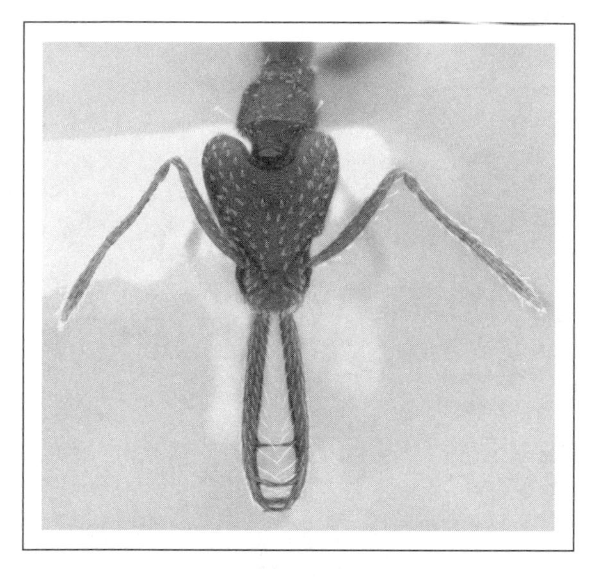

The face of a dacetine ant, *Strumigenys cordovensis.* Collected by Stefan Cover in Cuzco Amazonico, Peru. Imaged by Christian Rabeling.

Eleven

A Mentor and the Start of a Career

As a callow, severely undereducated eighteen-year-old student at the University of Alabama, I began a correspondence with a Ph.D. student at Harvard University named William L. Brown. Although only seven years my senior, Bill was already a leading world authority on ants. At that time there were only about a dozen experts on ants worldwide and he was one of them, not counting those who specialized on the control of pest species.

The most inspiring thing about Bill Brown was his devotion bordering on fanaticism—to science, to entomology, to jazz, to good writing, and to ants, in that rising order. He was, as I wrote of him in a 1997 memorial tribute, a working-class guy with a first-rate mind. He visited bars, enjoyed beer, dressed poorly by the stiff standards of the Harvard of his day, and mocked pretense whenever he encountered it in the faculty. But he was a godsend to the boy he befriended.

"Wilson," he wrote his teenage follower, "you've made a good start with your project of identifying all the species of ants found in Alabama. But it's time to get serious about a more basic subject, where you can do original work in biology. If you're going to study ants, get serious."

Bill, when I first came to know him, was at that time absorbed in classifying a group of species called the dacetine ants, limited mostly to the tropics and parts of the warm temperate zone. These insects are easily distinguished by their bizarre anatomy. Their jaws are long and hooked at the end and lined with needlelike teeth. Their bodies are clothed in various combinations of curly or paddle-shaped hairs; and, in many of the species, a spongy mass of tissue encircles the waist.

"Wilson," Bill went on, "there are a lot of species of dacetines in Alabama. I want you to collect as many colonies for our studies as you can, and while you're at it, find out something about their behavior. Almost nothing has been done on that subject. We don't even know what they eat."

I liked the way Bill Brown addressed me as a colleague, albeit one in training, like a sergeant instructing a private. If we had been in the U.S. Marines, I suppose I would have followed him to hell and back—or something like that, assuming there are ants living somewhere in hell. In spite of my young age and lack of experience, he expected me to behave as a professional entomologist. He insisted that I just get out there and get the job done. There was no hint of "get in touch with your feelings" or "think about what you'd most like to do."

So, pumped up with his confidence in me, I got out there and got the job done. I began by molding a series of plaster-of-Paris boxes with cavities the size of those that wild colonies occupy in nature. I added a larger adjacent cavity where the ants could hunt for prey. Into many such cavities I placed live mites, springtails, insect larvae, and a wide variety of other invertebrates I found around the nests of dacetines in natural habitats. I was later to label this the "cafeteria method."

My efforts were rewarded quickly. The little ants, I discovered, prefer soft-bodied springtails (technically, entomobryoid collembolans). As I watched them stalk and capture these prey, the odd anatomy of the dacetine ants made perfect sense. Springtails are abundant around the world in soil and leaf litter, and in some localities they are among the dominant insects. But ordinary predators such as ants, spiders, and ground beetles find them very difficult to catch. Beneath the body of each is a long lever that can be sprung violently but most of the time is locked in place—in other words, constructed like a mousetrap. When the springtail is disturbed even slightly, it pulls an anatomical trigger and the lever is released. Slamming against the ground, the lever catapults the insect high into the air. The equivalent acrobatic feat in a human being would be a leap of twenty yards up and a football-field distance forward.

The high jump works well against most predators, but the dacetine ant is built to defeat it. Upon sensing a springtail close by with the sensory receptors in her antennae—she is mostly blind—the huntress throws her long mandibles open, in some species 180 degrees or more, and locks them in place with a pair of movable catches on the front of the head. The huntress then slowly stalks the prey, literally step by cautious step. In the presence of a springtail, she is one of the slowest ants in the world. Her antennae wave side to side, also slowly, fixed on the location of the prey, turning to the right when the odor grows faint on the left, and to the left when the odor grows faint on the right, keeping the ant on track. Two long sensitive hairs project from the stalker's upper lip. When their tips touch the springtail, the catch is pulled down, releasing the powerful muscles that strain at the base. The mandibles slam shut, driving the needle-sharp teeth into the soft body of the springtail. Often the prey is able instantaneously to release its abdominal lever, throwing it and the ant spinning into the air. I've often thought that if dacetine ants and springtails were the size of lions and antelopes, they would be the joy of wildlife photographers.

From my and Bill Brown's early studies, various of which we published singly or together, a first picture of dacetine biology emerged. First, physiologists came to realize that the closing of the mandibles is one of the fastest movements that exist in the animal kingdom. Also the spongelike collar around the dacetine's waist was discovered by later researchers to be the source of a chemical that attracts springtails, drawing them closer to the mandibular snare.

In time we and other entomologists came to recognize the dacetines as among the most abundant and widely distributed of all ant groups. Although their tiny size makes them inconspicuous in the soil and litter, they are an important link of the food chains of the world's habitats. And, incidentally, colonies of many species live in rotting stumps like the one I described earlier.

During the next decade, Bill Brown and I took the next logical step into evolutionary biology. Armed with growing information, we reconstructed the changes in dacetines across millions of years, as they spread around the world and their species multiplied. In what manner and under what conditions, we asked, have the different species grown or shrunk in anatomical size? How and why did some of them evolve to build their nests in the soil and others in fallen twigs on the ground, or in rotting logs and stumps? A few, we learned, are even specialized to live in the root masses of orchids and other epiphytes of the rain forest canopy.

The history of the dacetine ants came into focus as we continued our studies. It turned out to be an evolutionary epic comparable to that of all the kinds of antelopes, for example, or all of the rodents, or all of the birds of prey. You may think that ants like these, being so small, must also be unimportant and deserving of less attention. Quite the contrary. Their vast numbers and combined weight more than make up for their puny individual size. In the Amazon rain forest, one of the world's strongholds of biological diversity and massed living tissue, ants alone weigh more than four times that of all the land-dwelling vertebrates—mammals, birds, reptiles, and amphibians—combined. In the Central and South American forests and grasslands alone, one taxonomic group of ants, the leafcutters, collect fragments of leaves and flowers on which they rear fungi for food, making them the leading consumers of vegetation. In the savannas and grasslands of Africa, mound-building termites also rear fungi and are the primary animal builders of the soil. Although insects, spiders, mites, centipedes, millipedes, scorpions, proturans, pillbugs, nematodes, annelid worms, and other such lilliputians are ordinarily overlooked, even by scientists, they are, nonetheless the "little things that run the world." If we were to disappear, the rest of life would flourish as a result. If on the other hand the little invertebrates on the land were to disappear, almost everything else would die, including most of humanity.

Because as a boy I dreamed of exploring jungles in order to net butterflies and turn over stones to look for different kinds of ants, I followed by happenstance the advice I gave you earlier: go where the least action is occurring. Just by any small twist of fate, I might easily have joined the large population of young biologists working on mice, birds, and other large animals. Like most of them, I would have enjoyed a productive and happy career in research and teaching. Nothing wrong with that at all, but by following the less conventional path, and by having an inspiring mentor like Bill Brown, I had a far easier time of it. I discovered early the special opportunity to conduct scientific research in rotting stumps and other microcosms that make up the foundation of the living world, but which then and to this day remain so easily passed by.

Martialis heureka, the most primitive known living ant. Modified from drawing by Barrett Klein, Biology Department, University of Wisconsin–La Crosse (www.pupating.org).

Twelve

The Grails of Field Biology

Tracking the history of the dacetine ants, Bill Brown and I came to focus on what appears to be the most primitive living species, similar to the ancestral species that long ago gave rise to the worldwide tribe of dacetines alive today. Our quarry was *Daceton armigerum*, a big insect as ants go, roughly the same size as the half-inch-long carpenter ants found everywhere in the north temperate zone. Covered with spines, its long jaws flat and armed at the tip by sharp spines, it was known to occur on trees in the rain forests of South America. Otherwise, entomologists had almost no information on where it nests, the social structure of its colonies, how and when it forages, and the kind of prey it hunts. It became, for a short while at least, my personal grail.

Very early in my ant-hunting world travels, I arrived in Suriname, at that time known as Dutch

Guiana. I went immediately into the rain forests around the capital city of Paramaribo to search for the big dacetine. After a week of sweat-soaked work and failure, I enlisted the help of resident entomologists. They in turn sent forth their assistants and a few other forest-savvy locals who had seen the ant and had a good idea where to look. Soon a colony was found. It was where I had not looked—in a small tree growing in a dense, seasonally flooded swamp. We cut the tree down and carried it in segments to a laboratory in Paramaribo. There I carefully and lovingly sliced open the trunk, revealing a cavity in which the entire colony lived—queen, workers, brood, and all. Studying it (and, later, a second colony I found in Trinidad), I filled in the blank spaces: The colonies are composed of several hundred workers; the foragers go out singly to search for prey in the canopy; each worker hunts on its own, catching insects of a wide variety, all of which are larger than springtails and other prey sought by smaller known dacetines. And more.

It is common for biologists to make a scan of biodiversity in order to locate some especially promising species or other, like the primitive giant dacetine, that offers opportunities to make a discovery of unusual importance. Another expedition I launched with the same goal in mind was to Ceylon, now known as Sri Lanka. The aneuretine ants found there I knew to be as distinctive a group as the dacetine ants. Unlike dacetines, however, aneuretines are not among the dominant insects of the world at the present time. In fact, they are on the edge of extinction. Their high moment in the evolutionary sweepstakes came long ago, toward the end of the Mesozoic Era, the Age of Reptiles, and continued on for a while into the early Cenozoic Era, the Age of Mammals—in other words a hundred million to fifty million years ago. We knew from fossil remains that aneuretines were both diverse and relatively common during the latter period. But of their social organization, their nests, their colonies, their communication, their food habits, we knew nothing. When I was a young researcher at Harvard, I was aware that in the late 1800s two specimens of a living species, *Aneuretus simoni*, had been collected in the six-hundred-year-old Royal Botanical Gardens in Peradeniya near Kandy, in the center of Sri Lanka. But no other specimens of the small dark-yellow ant had found their way into collections since that time.

Was the last living aneuretine species extinct? Had it gone the way of the dodo and Tasmanian wolf during such a brief interval of time, after tens of millions of years of life? I felt compelled to find out. Another grail! In 1955, at the age of twenty-five, I disembarked from an Italian passenger ship at Colombo and went straight to the Udawattakele, the forested pleasure garden of the kings at Kandy, which seemed to be the most promising semi-natural site. For a week I searched throughout the daylight hours. I came up with nothing, not even one stray aneuretine worker. I then proceeded to the more disturbed grounds of the Peradeniya gardens, the source of the original specimens. More close searching, still no *Aneuretus*. It seemed indeed possible that the species I sought, and with it the great evolutionary assemblage of the aneuretine ants, might really be gone.

But this verdict was unacceptable to me. So I traveled south to Ratnapura, resolved to hunt for the ant out from the city and into the nearby rain forest, which at that time stretched almost continuously to Adam's Peak.

Upon arriving in Ratnapura, I checked into a rest house, washed up, and within the hour strolled over to a nearby reservoir, where, although the shore was torn up by pedestrians and grazing cattle, I had noticed a thin grove of trees. I idly picked up a hollow twig lying on the ground and snapped it in two, expecting nothing much of interest to be living inside. Instead, I was stunned when out poured a stream of angry *Aneuretus*. I stood there staring at this wonderful gift. I paid no attention to the irritating sensation of the workers pouring over my hands. Would an Audubon scholar, in comparison, be bothered by a paper cut upon discovering a new original folio?

The next day, elated as I supposed only an

entomologist can be, I took a bus inland to a stop close to the edge of the nearby rain forest. I was accompanied by an assistant assigned to me by the Museum of Natural History in Colombo. His principal role was to assure local Jainists, whose religion forbids the killing of all animal life even down to the lowly ants, that I had been allowed a dispensation. Along a forest trail I soon found several more *Aneuretus* colonies. I studied them in the field, during intervals between occasional pounding downpours of rain. Several colonies I placed in artificial nests to study their communication, care of the young and mother queen, and other aspects of their social behavior. Back at Harvard, I worked with several colleagues to describe the aneuretine internal anatomy.

Almost thirty years later, as a Harvard professor, I directed an undergraduate student from Sri Lanka, Anula Jayasuriya, as she made further surveys of the aneuretines for her senior honors thesis. She found that the range of the species was shrinking, which was no surprise due to the relentless clearing of Sri Lanka's lowland forest since the time of my visit. At this point I had *Aneuretus simoni* put on the list of endangered species compiled by the International Union for the Conservation of Nature, one of the few rare insect species well enough known even to be considered for this category.

During this period, the picture of the evolution of the small but world-dominant ants as a whole was coming into focus. More researchers were entering the study of fossil and living species. We were filling in the steps in evolution that led to surviving groups, while discovering previously unknown groups and the ancestral lines that linked them together.

For a while the largest gap of all remained, the ancestor of all the ants. There is no such thing as a living solitary ant. All living species, so far as we know, form colonies with a queen and her sterile (or almost sterile) daughters, who do all the work. Males are raised in the nest solely for the purpose of mating with virgin queens. They leave the nest to find mates, are not allowed to return, and soon die. King Solomon, who instructed, "Go to the ant, thou sluggard, consider her ways, and be wise," was obviously not aware of all the facts of ant biology in his moral urging. Nonetheless, how did this bizarre but extremely successful social system come into existence? When I was a young scientist we had many fossils to study, some dating back to more than fifty million years before the present, but every species represented had worker castes. Of the origin of their social organization we knew nothing.

This grail we ant biologists sought was a link still missing—a primitive ant with colonies like those of the ancestral forms that lived more than fifty million years ago, and simple enough to provide clues to the origin of social behavior. The leading candidate of which we had knowledge at this time was the Australian dawn ant (*Nothomyrmecia macrops*). Unfortunately, like the living aneuretines of Sri Lanka, the species was known from only two specimens. These had been collected in 1931 in one of the most remote places in the world. The land was the relatively inaccessible sand-plain heath of Western Australia. In the 1950s this vast area, stretching from the small coastal town of Esperance in the west to the edge of the desertlike Nullarbor Plain in the east, and covering over ten thousand square miles in area, was entirely devoid of people. Two decades before my own visit, a party of adventurers had traveled on horseback through this heath from the transcontinental highway south to an abandoned homestead on the coast called the Thomas River Farm, thence about a hundred miles west to Esperance. The terrain they crossed is one of the biologically richest in the world. In the seemingly barren scrubland lived large numbers of plant species found nowhere else on Earth. The insects were mostly unknown to science.

With the group in 1931 was a young woman who had agreed to collect ants along the trail for John S. Clark, an entomologist at the Museum Victoria in Melbourne and the sole expert on ants in Australia at that time. The collector carried a jar of alcohol into which she dropped ants wherever she found them.

When Clark examined the specimens he was startled to find two belonging to a previously unknown ant species, primitively wasplike in form. It appears to be closest in anatomy among all known living ants to what may have been the ancestor of all ants. Unfortunately, the collector kept no records during the trek of where particular ant species had been found. The Australian dawn ant might have been picked up anywhere along a hundred-mile line.

By the time I arrived in 1955 to study Australian ants, I was obsessed with the idea of rediscovering this enigmatic species. It was already a legend among naturalists. I wanted to know whether it was fully social, with well-organized colonies of queens and workers, or less so—perhaps just partway to the advanced condition of all other known ants. Biologists of the time otherwise had no idea of how advanced ant social life had originated, or why.

Still young at twenty-five and charged with energy and optimism, I invited two fellow enthusiasts to join me in the effort to rediscover *Nothomyrmecia macrops*. One was Vincent Serventy, a famous Australian naturalist and authority on the Western Australian environment. The other was Caryl Haskins, a longtime ant expert and at that time the newly appointed president of the Carnegie Institute of Washington. We rendezvoused in Esperance, loaded up on supplies, and headed east in an old army flatbed truck along a dirt track to the Thomas

River Farm. The flat plain, clothed in flowering shrubs and herbaceous plants, was beautiful to behold and blessedly empty—we saw only one other vehicle during the entire trip. From this base we searched outward in all directions, night and day, for the better part of a week. Dingoes prowled around our camp at night, the summer sun dehydrated us, and our footsteps turned huge meat ant nests into boiling masses of angry red-and-brown, viciously biting defenders. Was I afraid? Never. I loved every minute of it.

We devoted one day of our search to a trip northward to Mount Ragged, a prominence on whose barren sandstone slopes the dawn ants might have been collected. The only water source, for the 1931 party and ourselves, was a moist spot on the roof of a shaded ledge, from which enough water dripped to fill one cup each hour. No dawn ants were located there either.

Our overall effort yielded many new species of ants, but not a single specimen of the dawn ant. Because of my high expectations, the failure was one of the greatest disappointments of my scientific life.

Our failed expedition was nevertheless widely publicized in the Australian press, and it stimulated further searches in the sand-plain heath by entomologists. There was a widespread feeling among the local scientific cognoscenti that if this special insect was to be rediscovered and studied, it should be by Australians and not by Americans, of whom more than enough had already visited the continent.

One such attempt was led by my former student Robert W. Taylor, who had completed his Ph.D. at Harvard and at the time was a curator of entomology at the national insect collections in Canberra, the capital of Australia. Bob was desperate to make the discovery, to seize this grail for himself and for the honor of Australian entomology. On the way west to dawn ant country, the group camped in a forest of mallee, a kind of shrubby eucalyptus. The night was chilly, and there seemed to be no good reason to search for any insects at all. But Taylor walked out anyway with flashlight in hand, just in case something might be active. A few minutes later he came running back, shouting, "I got the bloody bastard! I got the bloody bastard!" As his words hint, now famous among entomologists, the dawn ant had indeed been found—and if not by an Australian, at least by a New Zealander.

It turned out that the dawn ant is a winter species. The workers wait in their nests and come out on cool nights to forage for mostly insects, many of which are numbed and easy to catch. The species is part of the ancient Gondwanan fauna, insects and other creatures of which a large part originated in Mesozoic times during the early breakup of the Gondwanan supercontinent and the drift northward of New Zealand, New Caledonia, and Australia. The relict elements, of which the dawn ant is part, are species adapted to the south temperate zone, and sometimes to the cool-temperature regimes of winter. I should have anticipated that possibility when searching in midsummer out of Esperance. But I didn't.

With a population of dawn ants located, a flood of studies followed, during which virtually every aspect of the biology and natural history of the species was explored. Dawn ants proved to be elementary in most aspects of their social behavior, but they are not the fundamentally less social creatures we had hoped to find. Like all other known ants, they form colonies with queens and workers. They build nests, forage for food, and raise their sisters. All are cooperating subordinate daughters of the mother queen.

To discover the origin of all the ants, even taking into account their diminutive stature, is as important as finding the origin of dinosaurs, birds, and even our own distant ancestors among the mammals. I realized that without a satisfactory living link, researchers needed to find the right fossils from the right geological period to make further progress. Until 1966, however, the earliest known fossils were between a relatively youthful fifty million and sixty million years old, by which time, in the early to middle Eocene Period, the ants were already abundant and highly diversified. They were also globally distributed. We had even found an extinct species of dawn ant similar to the living one of Australia, preserved in the Baltic amber of Europe.

It was all very frustrating. Ants obviously had arisen during the Mesozoic Era, which ended sixty-five million years ago. But for a long time we had not a single Mesozoic specimen. It seemed as though a dark curtain had been lowered over the ancestors and earliest species of these world-dominant insects. Then, in 1966, word came to Harvard that two specimens of what appeared to be ants had been found in ninety-million-year-old amber from a geological deposit in, of all places, not some exotic far-off fossil bed but smack on the shores of New Jersey, and they were on the way for me to examine. At last the curtain might lift! I was so excited that when I fished the amber piece out of the mailing package I fumbled and it dropped to the floor. It broke into two pieces that skittered away from each other. I was aghast. What disaster had I wrought? However, to my great relief each piece contained an entire separate ant, and neither of the fossils had been damaged. When I polished the surface of the pieces into glassy smoothness, I found the external form of the specimens to be preserved almost as though they had been set in resin only a few days earlier.

My collaborators and I named the Mesozoic ant *Sphecomyrma freyi*, the first generic name meaning "wasp ant," and the second in honor of the retired couple who had found the specimens. The generic name was fully justified: the species had a head that was mostly wasplike, some parts of the body were mostly antlike, and other parts of the body were intermediate in form between wasps and ants. In short, the missing link had been discovered, another grail found.

The announcement of the discovery set off a flurry of new searches by entomologists for ants and antlike wasps in amber and sedimentary rock deposits of late Mesozoic age. Within two decades many more specimens turned up in deposits from New Jersey, Alberta, Burma, and Siberia. In addition to more *Sphecomyrma*, new species at other levels of evolutionary development came to light. The story of the early diversification of the ants began to unfold. We found that it reaches back at least 110 million years and probably well beyond, to as far as 150 million years before the present.

Yet, sadly, we still had only fossils. No living evolutionary links had been found whose social behavior could be studied in the field and laboratory. It appeared that direct knowledge of the early stages of social behavior in the ants might have to be pieced together indirectly. The Australian dawn ant and a small number of other comparably primitive lines among the living ants might prove the best that would ever be found.

Then in 2009 came a complete surprise with

at least the potential to change the big picture. A young German entomologist, Christian Rabeling, was excavating soil and leaf litter in rain forest near Manaus, in the central Amazon. Rabeling, with whom I've since worked in the field, has the deserved reputation of leaving, literally, no stone unturned. He also readily climbed trees, unaided by equipment, to bring down ant colonies nesting in the canopy. One day, as he was picking up every new kind of ant he could find, he spotted a single pale, odd-looking specimen crawling beneath the fallen leaves. Picking it up, he realized that he could not place it to any known genus or species of ants.

During a visit to Harvard he brought his discovery along with the rest of his collection to the "Ant Room." Here, in cramped quarters on the fourth floor of Harvard's Museum of Comparative Zoology, is kept the largest and most nearly complete classified collection of ants in the world. Built up by a succession of entomologists over more than a century, it contains perhaps a million specimens (no one has volunteered to make an exact count), belonging to as many as six thousand species. Ant experts from around the world come to these quarters to identify specimens they have collected on their own, and to conduct research on classification and evolution. Several were present when Rabeling brought in his Amazonian oddity.

After much consternation, the group invited me in from my office across the hall. I remember the moment vividly. Taking a look under the microscope, I said, "Good God, this thing must be from Mars!" Which meant I didn't have a clue either. Later, when Rabeling described the species formally in a technical journal, he gave his ant the name *Martialis heureka*, which means, roughly, "the little Martian that has been discovered." It was an ant, all right, and proved an earlier branch in the ant family tree than even the Australian dawn ant. At this writing three years later, no further Martialis ants have been found. The Amazon is a very big place to look, however, and I expect a colony will eventually be located if the species is truly social, and perhaps by one or more of the growing group of young ant experts in Brazil.

You may think of my story of ants as only a narrow slice of science, of interest chiefly to the researchers focused on it. You would be quite right. But it is nonetheless at a different level from an equally impassioned devotion to, say, fly fishing, Civil War battlegrounds, or Roman coins. The findings of its lesser grails are a permanent addition to knowledge of the real world. They can be linked to other bodies of knowledge, and often the resulting networks of understanding lead to major advances in the overall epic of science.

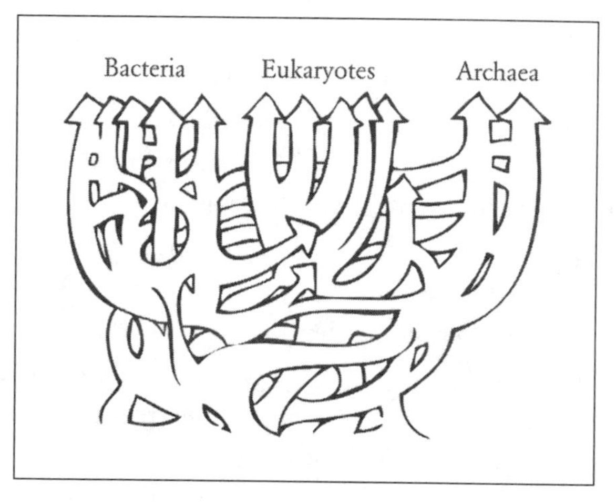

The basic tree of life with gene exchanges during the earliest evolution, as envisioned by the microbiologist W. Ford Doolittle. Modified from the original drawing in "Phylogenetic classification and the universal tree," by W. Ford Doolittle, *Science* 284: 2124–2128 (1999).

Thirteen

A Celebration of Audacity

Six years before the discovery of the archetypical ant *Martialis* in the Amazon forest, a major effort had begun by entomologists to work out the family tree, more technically called the branching phylogeny, of all the living ants. Therein lies yet another chapter of my story especially relevant to you. In 1997 I had finally retired from the Harvard faculty and stopped accepting new Ph.D. students. Nevertheless, in 2003, the chairman of the Graduate Committee of the Department of Organismic and Evolutionary Biology called one day and said to me, "Ed, we've already accepted our quota of new students for this year, but we've got one more, a young woman so unusual and promising that we'll add her on if you'll agree to be her de facto sponsor and supervisor. She's a fanatic on ants, wants to study them above all else. And she has tattoos of ants on her body to prove it."

Dedication like that I admire, and after looking at her record I saw that Harvard was ideal for her. And she, it seemed, would be ideal for Harvard. I recommended that Corrie Saux (later Corrie Saux Moreau) from New Orleans be forthrightly admitted. When she appeared I knew we had made the right decision. She breezed through the first-year basic requirements. By the end of the year she already had a clear idea of what she wished to do for her Ph.D. thesis. Three leading experts on ant classification, each in different research institutions, had just received a multimillion-dollar federal grant to construct a family tree of all the major groups of ants in the world, based on DNA sequencing—the ultimate technique for the job. It was an important but formidable undertaking that, if successful, would undergird studies on the classification, ecology, and other biological investigations of all of the world's sixteen thousand known ant species. Also, understanding the ants, many of the specialists realized, means learning a great deal more about Earth's terrestrial ecosystems.

Saux suggested that she write the three lead researchers for permission to decode one of the smaller taxonomic divisions of the ants (one out of the twenty-one in all). I said, yes, it would be an achievement worth a degree if she could manage it, and a good way to meet other experts and work with them.

Soon afterward, however, she came back to tell me that the project leaders had turned her down. They were disinclined to add a new, untested graduate student to the team. From my own student days, I had learned to have a tough skin, not to accept a no as a personal rejection. With that in mind, I said, "Okay, don't let that get you down. What the project leaders decided isn't a bad thing. Why don't you pick something else that you'd like to do?"

A few days later she came back and said, "Professor Wilson, I've been thinking, and I believe I could do the whole project myself." I said, "The whole project?" She responded with demure sincerity, "Yes, all twenty-one of the subfamilies, all the ants. I think I can do it."

Corrie then added that the world-class collection at Harvard was a great advantage. All she needed, she said, was a postdoctoral assistant who had specialized in DNA sequencing. She knew one who was willing to take the job. Might I supply the money for his salary? After a pause, I said impulsively, more out of instinct than logical reflection, "Well, okay."

There was no bravado in Corrie, no trace of overweening pride, no pretension. She was a quiet, serene enthusiast. As it turned out, she was also an open, helpful friend to fellow students and others around her. She'd come from New Orleans by way of San Francisco State University, and I took pride in her as a fellow southerner. I wanted her to succeed, and while I did not join as a collaborator, I found the funds to set up her laboratory. And why not? An effort like this celebrates imagination, hope, and audacity. And there was a fallback position for Corrie: if she fell short of the whole, she could use the part completed as a thesis. I even helped, a little, on the side. When I visited the Florida Keys on another project during the months that followed, I collected live ants of the genus *Xenomyrmex* for her, filling in a group difficult to obtain in the field. Along the way, she told me she needed to consult with an expert on some complex methods in statistical inference. I funded that also.

At this point I was determined to see Corrie Saux to the end. I felt that she could actually accomplish what she envisioned.

Her thesis was finished in 2007, read closely by her Ph.D. committee, and approved. On April 7, 2006, the core of her study was published as the cover article in *Science*, an achievement that would be considered exceptional even for a senior researcher. I admit I was nevertheless a bit tense when Corrie's thesis went to the Harvard committee for review.

Then I learned that the three-person team with the larger grant had also finished their work and planned to publish the results later in the year, allowing history to record that the two studies had been conducted independently and simultaneously. Of this I warmly approved, especially since each of the three was a highly regarded scientist. But it also meant that Corrie Saux's research was about to be thoroughly tested. What if the two phylogenies didn't match? That was a scenario I didn't want to think about.

To my great relief, however, the two phylogenies matched almost perfectly. There was a difference in the placement of one of the twenty-one subfamilies, the leptanilline ants, an obscure and little-known group. Even that variance in interpretation was later worked out through more data and statistical analysis.

The story of Corrie Saux Moreau's ambitious undertaking is one I feel especially important to bring to you. It suggests that courage in science born of self-confidence (without arrogance!), a willingness to take a risk but with resilience, a lack of fear of authority, a set of mind that prepares you to take a new direction if thwarted, are of great value—win or lose. One of my favorite maxims is from Floyd Patterson, the light heavyweight boxer who defeated heavier men to win and for a while hold the heavyweight championship. "You try the impossible to achieve the unusual."

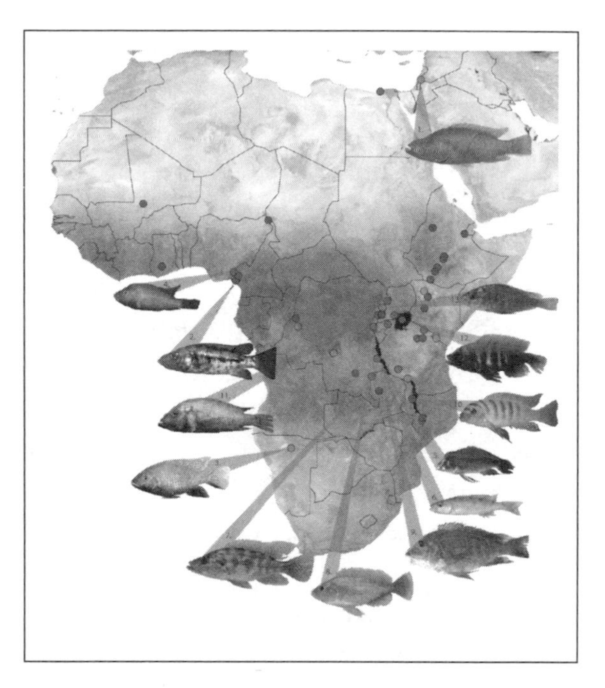

Locations of the evolution of cichlid fish species in Africa. Modified from "Ecological opportunity and sexual selection together predict adaptive radiation," by Catherine E. Wagner, Luke J. Harmon, and Ole Seehausen, *Nature* 487: 366–369 (2012). doi:10.1038/nature11144.

Fourteen

Know Your Subject, Thoroughly

To make discoveries in science, both small and important, you must be an expert on the topics addressed. To be an expert innovator requires commitment. Commitment to a subject implies sustained hard work.

If you look beneath the surface of important discoveries to obtain a glimpse of the scientists who made them, you will soon see the truth of this generalization. Here, for example, is testimony from the theoretical physicist Steven Weinberg, who with Sheldon Lee Glashow and Abdus Salam won the 1979 Nobel Prize in physics for "contributions to the theory of the unified weak and electromagnetic interaction between elementary particles, including, interalia, the prediction of the weak neutral current":

> I was born in New York City to Frederick and Eva Weinberg. My early inclination toward science received encouragement from my father, and by the time I was 15 or 16 my interests had focused on theoretical physics . . .
>
> After receiving my Ph.D. in 1957, I worked at Columbia and then from 1959 to 1966 at Berkeley. My research during this period was on a wide variety of topics—high energy behavior of Feynman graphs, second-class weak interaction currents, broken symmetries, scattering theory, muon physics, etc.—topics chosen in many cases because I was trying to teach myself some area of physics. My active interest in astrophysics dates from 1961–62; I wrote some papers on the cosmic population of neutrinos and then began to write a book, *Gravitation and Cosmology*, which was eventually completed in 1971. Late in 1965 I began my work on current algebra and the application to the strong interactions of the idea of spontaneous symmetry breaking.

Obviously, Steven Weinberg did not just wake up one morning, reach for pencil and paper, and sketch out his breakthrough insights.

Switching to a very different subject, X-ray crystallography, we have James D. Watson's characterization of Max Perutz and Lawrence Bragg. It is in *The Double Helix*, arguably the best memoir ever written by a scientist, a book I recommend to any young person who wants to experience almost personally the thrill of scientific discovery. In it he describes what proved to be the essential step for solving the structure of the all-important coding molecule:

> Leading the unit to which Francis [Crick] belonged was Max Perutz, an Austrian-born chemist who came to England in 1936. [Perutz] had been collecting X-ray diffraction data from hemoglobin crystals for over ten years and was just beginning to get somewhere. Helping him was Sir Lawrence Bragg, the director of the Cavendish. For almost forty years Bragg, a Nobel Prize winner and one of the founders of crystallography, had been watching X-ray diffraction methods solve structures of ever-increasing difficulty. The more complex the molecule, the happier Bragg became when a new method allowed its elucidation. Thus in the immediate postwar years he was especially keen about the possibility of solving the structures of proteins, the most complicated of all molecules. Often, when administrative duties permitted, he visited Perutz' office to discuss recently accumulated X-ray data. Then he would return home to see if he could interpret them.

During nearly two decades, from 1985 to 2003, I brought to reality a dream that others before me considered inordinately difficult or even impossible. Fitted in between my classes at Harvard in the years before I retired, as well as other research and writing projects, I undertook the classification and natural history of the gigantic ant genus *Pheidole*. This is no ordinary group. It comprises by far the largest number of species of any ant genus, and further, it is among the largest genera of animals and plants of any kind. In many regions of the world, from desert to grassland to deep rain forest, it is also frequently the most abundant of all ants. What distinguishes *Pheidole* is the possession of two castes, slender minor workers and much larger big-headed soldiers. The possession of such variation within colonies adds to the biological complexity of these remarkable insects.

So great was the species roster that the taxonomy of *Pheidole* when I started my revision was in a shambles. Most of the species recognized by earlier classifiers were unrecognizable from the brief descriptions given them. Worse, the collections of specimens accumulated over the previous century were scattered among half a dozen museums in the United States, Europe, and Latin America. By the time I picked up the task, *Pheidole* could no longer be ignored. Its many species are collectively among the major players in the environment. Ecologists trying to understand symbioses, energy flows, the turning of soil, and other basic phenomena were unable to name the species they were observing. Except for collection sites in North America, they were usually forced to report their specimens as belonging to "*Pheidole* species 1, *Pheidole* species 2, *Pheidole* species 3," and so on to species 20 and beyond. This might work, at least roughly, for one researcher at one locality. But other biologists at other localities had their own independent rosters. Their *Pheidole* species 1, species 2, species 3, and so forth were by chance alone very likely different from the rosters of others, and the lists could be collated only if the researchers undertook the tedious task of bringing the specimens together. Better if from the start all writers used the same comprehensive list, comprising, for example, *Pheidole angulifera*, *Pheidole dossena*, *Pheidole scalaris*, and so on, each species having been defined earlier in a careful, formal manner and made universally convenient in the literature. When the taxonomy has been straightened out, biologists wishing to study the genus could identify the species to their single acceptable name. They could immediately collate their findings with those of other researchers, and pull from the literature everything previously known about every species of interest.

Taxonomy is often spoken of as an old-fashioned discipline. Some of my friends in molecular biology used to call it stamp collecting. (Maybe some still do.) But it is emphatically not stamp collecting. Taxonomy, or systematics, as it is often called to spiff up its image, is fundamental to modern biology. In technology it is conducted with the aid of sophisticated field and laboratory research, using DNA sequencing, statistical analyses, and advanced information technologies. To take its place in basic biology, it is grounded in studies of phylogeny (the reconstruction of family trees) and in analyses of the genetics and geographical research devoted to the multiplication of species. The task of taxonomy drawing from these disciplines is made the more difficult, however, by the fact that most species of animals and microorganisms, together with a substantial minority of plants, await discovery.

Ant taxonomists called the genus *Pheidole* the Mount Everest of ant taxonomy, towering arrogantly in front of us, seemingly too big to be mastered. There were many lesser but still important challenges on which others could build a productive career. I could face failure, I thought, so I took the job of ascending the ant Everest, at first in collaboration with my old mentor William L. Brown. When Bill's health began to decline soon afterward, I soldiered on the rest of the way, starting with the Western Hemisphere, the biodiversity headquarters of the genus. I felt obligated to continue to the end, in part because I was located at the Museum of Comparative Zoology, with easy access to the largest collection and best library in the world suited to the task. But I also persisted partly for the challenge and partly because I thought of it as my duty. In the end, when *Pheidole in the New World: A Dominant, Hyperdiverse Ant Genus* was published in 2003, the book comprised 798 pages in which 624 species were diagnosed, 334 of them new to science, with everything known of the biology of every species cited, and all of the species illustrated, with a total of over 5,000 drawings I had made myself. Even as copies of *Pheidole in the New World* were being printed, new species continued to pour into the museum from collaborators in the field. It is likely that by the end of the century the total number of species will exceed 1,000, perhaps even 1,500, species.

I planted our flag on the Pheidole summit, so to speak, but I am no Edmund Hillary or Tenzing Norgay. I had another goal in mind while encompassing the classification of the monster genus. One was to discover new phenomena in the course of giving thought to each species in turn. I was following the second of two strategies I gave you in an earlier letter: *for each kind of organism there exists a problem for the solution of which the organism is ideally suited.* One success in this correlative effort was the discovery of the "enemy specification" phenomenon.

The principle behind its concept is simple. Every species of plant and animal is surrounded in its natural habitat by other species of plants and animals. Most are neutral in their effect upon it. A few are friendly, and at the extreme, there is the symbiotic level. In the latter case, two or more are dependent upon one another for their very survival or at least reproduction—for example, pollinator animals and the plants they pollinate. A few other plant and animal species are, on the other hand, inimical to a particular species, so much so in a few cases as to be dangerous to their survival. It is to the great advantage of individuals of that species to recognize dangerous enemies instinctively and to avoid or destroy them if possible.

The principle sounds like common sense. But do species really evolve such an enemy specification response? I had never thought of it much one way or the other. Instead, I discovered it by accident. During the *Pheidole* project I cultured laboratory colonies of *Pheidole dentata*, an abundant species through the southern United States. I also kept colonies of fire ants (*Solenopsis invicta*). One day I was conducting one of my easy, quick experiments by placing other kinds of ants and insects next to the artificial nest entrances of the *Pheidole dentata* colonies just to see how they would respond. I was especially curious to see which ones would draw out the powerful big-headed soldiers.

The response was usually tepid. Either the ants contacting the intruder retreated into the nest or, with a few other nestmates, engaged it in combat. But when I dropped just a single fire ant worker at the same spot, the reaction of the colony was explosive. The first forager to encounter the intruder rushed back into the nest, laying an odor trail as it ran, while frantically contacting one nestmate after the other. Both minor workers and soldiers then poured out of the nest, zigzagging and circling in a search for the fire ant worker. When they found it they attacked it viciously. The minor workers bit and pulled its legs, while the soldiers, employing their sharp mandibles and powerful adductor muscles that fill their swollen heads, simply chopped off the appendages of the fire ant to render it helpless.

The fire ants are certainly enemies of the deadly kind. When, in the laboratory, I placed *Pheidole* and fire ant colonies close together, some of the fire ant scouts made it back home alive to report their find and recruit nestmates to the battle. The far larger fire ant colonies quickly destroyed and ate their opponents. Yet in some natural habitats, colonies of both species are abundant. It became apparent that *Pheidole* survive by building their nests a safe distance from the fire ant colonies and killing off fire ant scouts before they can report home.

Later, in the Costa Rican rain forest, I found an even more remarkable response by another species (*Pheidole cephalica*) to rain or rising water that threatens to flood their nests. When I placed as little as a drop or two at the entrance of a nest, minor workers quickly mobilized the colony, and the whole emigrated within minutes to another location.

Discoveries like these, whether minor or important—and who is to say at first which it will be?—can be made only rarely without a thorough advance knowledge of the organisms studied. This precondition is sometimes called "a feel for the organism."

Let me relate another story to reinforce this important principle. It occurred during an expedition I led in 2011 to the South Pacific. With me were Christian Rabeling, the ant expert and discoverer of the Amazonian "Martian" ant; Lloyd Davis, another ant expert and world-class birder; and Kathleen Horton, who was in charge of the complex logistics. We traveled during the austral spring of November and early December. Our destination was two archipelagoes, the independent island nation of Vanuatu and the nearby French possession of New Caledonia. In the process we visited localities where I had collected and studied ants in 1954 and

1955. I looked forward to observing changes in the environment that undoubtedly had occurred fifty-seven years later. I brought scanned images of my aging Kodachrome slides with me to make the comparisons exact. In particular I wanted to evaluate the condition of the wildlands and the reserves and national parks since 1955.

What original discoveries we made, in particular with the ants we planned to collect and study, would depend entirely on the knowledge we brought with us. We were in fact well prepared. We discovered many new species, and kept notes on the habitats in which they were found. But that was only part of the plan. We had bigger game in mind: to clarify, if we could, phenomena in the formation of species and their spread from one island group to another across the intervening ocean gaps. If you look at a map of the South Pacific and make Vanuatu your focus, you see how plants and animals that colonized this archipelago could have come from any of three bodies of land: Australia and New Caledonia to the west, the Solomon Islands to the north, Fiji to the east, or some combination of all three. Ant colonists, although completely landbound, might have made the journey by floating on the logs and branches of fallen trees or blown by storm winds. Queen ants capable of founding colonies might even have ridden in the feathers of far-ranging birds. We could not hope to determine how ants cross open water, but we did collect enough data to judge which island group contributed the most colonies to Vanuatu. It turned out, incidentally, to be the Solomon Islands.

This discovery was important enough to justify the hard work in the field, but we devised another question to ask and perhaps answer. Leaving aside the Solomon Islands, whose ant fauna was still poorly explored, we were aware of a huge difference between Vanuatu and the two archipelagoes on either side of it, Fiji and New Caledonia. Both are ancient, having existed with a substantial land area for tens of millions of years. Vanuatu has been in existence for a comparable period of time, but only as a set of small, shifting islands. Only during the last million years has its land area been more than a tenth of what it is today. The antiquity of Fiji and New Caledonia is immediately apparent in the richness of their faunas and floras. In particular, each is occupied by a large number of species, some highly evolved, that occur nowhere else in the world.

And what of relatively youthful Vanuatu? In November 2011 we were the first to take a close look at the ants on this archipelago. We knew that if it had a long geological history and large land area like New Caledonia and Fiji, we should expect to find a rich, highly evolved array of ants present. If, on the other hand, the current large area of Vanuatu had a relatively short history, as the geologists claimed, we should find a much sparser, distinctive array of ants there than occur on Fiji and New Caledonia. As it turned out, we found a smaller array, in accord with expectations from the record deduced by the geologists. But the ants of Vanuatu have not been inactive during their "brief" million-year tenure. We found clear evidence of new species in formation, and the beginning of the kind of expansion of biological diversity that is well advanced on the older archipelagoes. The ants of Vanuatu, to put the matter as succinctly as possible, are in the springtime of their evolution.

I have one more story to tell you from the South Pacific, because it is about a process unfolding there that may at first appear to be remote and exotic yet has global significance. It makes urgent the lesson of knowing where you are and what to look for when doing field research.

While on New Caledonia, our little team joined Hervé Jourdan, a seasoned resident entomologist of the local Institute of Research and Development. He led us on a trip to the Isle of Pines, a small island off the southern tip of the main island, Grand Terre, and, at least from the viewpoint of Americans, one of the remotest places in the world. Our goal was to learn what kinds of ants occur there, and to search for one species in particular, the bull ant

Myrmecia apicalis. Bull ants are evolutionary cousins of the Australian dawn ant, and almost as primitive as that species in anatomy and behavior. Eighty-nine species of *Myrmecia* have been discovered in modern-day Australia. Only one, *Myrmecia apicalis*, is native elsewhere. The existence of this insect so far from the homeland raised questions of interest to biogeographers, whose business is to map and explain the distribution of plants and animals. When and how did the New Caledonian bull ant get to this remote archipelago? Which of the eighty-nine species back home in Australia are its closest relations? How has it adapted to the island environment? In which ways, if any, has it become special?

I wanted very much to answer these questions when I visited New Caledonia in 1955, but I could not find the species at all. The forest where it had been last seen on Grande Terre, the main island of the New Caledonian archipelago, had been cut over in 1940. In later years *Myrmecia apicalis* was considered extinct. But then Hervé Jourdan found several workers of the ant in a forested area on the Isle of Pines. We went there with him to locate colonies if possible and to learn all we could about this endangered species. To our relief we succeeded in finding three nests deep in undisturbed forest, and were able to film and study the ants day and night. The nests were located at the bases of small trees. Their hidden tunnels were capped with debris. Foraging workers, we found, leave the nest at dawn, climb singly into the canopy, return bearing caterpillars and other insect prey at dusk. Later we learned that *Myrmecia apicalis* is most closely related to a few Australian bull ant species with similar habits that live in the tropical forests of northeastern Australia. We still do not know how one such species was able to colonize New Caledonia, or how many thousands or millions of years ago it made the trip.

I'm telling you this faraway bit of natural history for a special reason. While on the Isle of Pines we confirmed the existence of a frightening threat to a large part of the island's biodiversity, not just the New Caledonian bull ant, but a large part of the fauna. Another ant, accidentally introduced to New Caledonia in cargo in recent years, has reached the small offshore island of Isle of Pines and is taking over the forests there, destroying, as it spreads, the native ants, other insects, and in fact almost all of the ground-dwelling invertebrates.

The alien enemy is the "little fire ant" (technical name: *Wasmannia auropunctata*), which originated in the forests of South America. With humanity's unintended help, the species is spreading throughout tropical regions of the world. I had first encountered this alien in the 1950s and 1960s in Puerto Rico and the Florida Keys. Since then it has reached and begun to expand in New Caledonia, where it is an especially destructive pest. Although its workers are tiny, the colonies are huge and aggressive. The species is as bad as the more famous imported fire ant (*Solenopsis invicta*), which has spread widely in warm temperate countries. The government of neighboring Vanuatu, aware of the dangers posed by the Wasmannia, is attempting to keep it at bay by spraying and exterminating beachhead populations whenever they are found on the islands.

The little fire ant is a particularly severe menace on the Isle of Pines. During our search for the bull ants and other entomological treasures, we visited several types of forests, including those composed of nearly pure stands of *Araucaria*, one of the signature plants of the New Caledonian archipelago. These towering steeple-shaped trees have prevailed on the fringes of the southern continents for tens of millions of years. We found that where the little fire ants had penetrated *Araucaria* groves, native ants and other invertebrates were almost entirely absent. The New Caledonian bull ants survived in a Wasmannia-free area, but that was only a mile or two from the slowly advancing fire ant wave. The final extinction of these unique insects, and very likely other native animals, might be only decades away.

Can the little fire ants be stopped? The French scientists at the Institute of Research for Development in Noumea have tried to find a way, but so far have met only failure. You may be thinking at this point that if Grande Terre and the Isle of Pines are so far away, why should we be concerned? I will answer with emphasis: because the little fire ants are only one of thousands of similar aliens spreading around the world. The number of invasive species of plants and animals, including disease-carrying mosquitoes and flies, home-destroying termites, pasture-choking weeds, and enemies of native faunas and floras, is increasing exponentially in every country. Invasive species are the second most important cause of extinctions of native species, exceeded only by the destruction of habitats through human activity.

To learn more of the details of the great invasive threat, and to find solutions before it has reached catastrophic levels, will require far more science and science-based technology than we now possess. Humanity needs more experts who have the passion and breadth of knowledge to know what to look for in the first place. That's where you come in, and why I have told you this story of New Caledonia's threatened bull ant.

IV

THEORY

and

THE BIG PICTURE

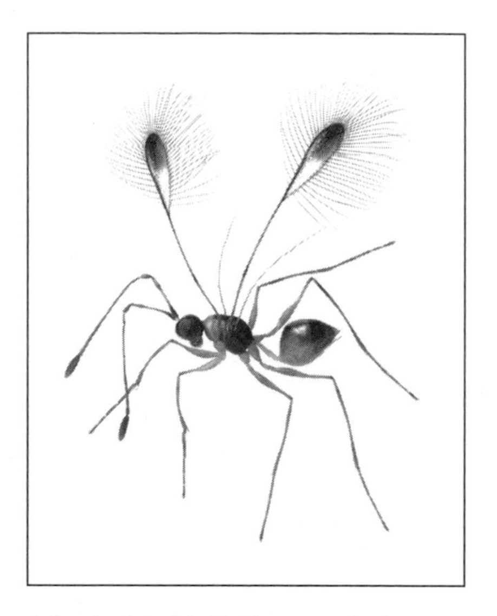

A female of the fairyfly *Mymar taprobanicum*, a wasp parasite of insect eggs. The actual size is smaller than the first letter in this caption. © Klaus Bolte.

Fifteen

Science as Universal Knowledge

There is only one way to understand the universe and all within it, however imperfectly, and that is through science. You are likely to respond, Not true, there are also the social sciences and humanities. I know that, of course, I've heard it a hundred times, and I've always listened carefully. But how different at their foundations are the natural sciences, social sciences, and humanities? The social sciences are converging generation by generation of scholars with biology, by sharing methods and ideas, and thereby conceding more and more to the realities of the ultimately biological nature of our species. Granted that many in the humanities, as if in a bunker, fiercely defend their isolation. Moral reasoning, aesthetics, and especially the creative arts are forged independently from the scientific world view. The stories of human relationships in history and the creative arts are potentially infinite, like music played upon only a few musical instruments. Yet however much the humanities enrich our lives, however definitively they defend what it means to be human, they also limit thought to that which is human, and in this one important sense they are trapped within a box. Why else is it so difficult even to imagine the possible nature and content of extraterrestrial intelligence?

Speculations about other kinds of mind are not pure fantasy. Rather, if informed they are thought experiments. Let's try one. Imagine with me that termites had evolved a large enough size to have brains with a capacity equal to that of humans. That may sound entirely implausible to you. Insects have exoskeletons that encase their bodies like a knight's armor. They cannot grow to be much larger than a mouse—and a human brain by itself is bigger than a mouse. But wait! Allow me a bit of flexibility in the scenario. In the Carboniferous Period on this planet, 360 to 300 million years ago, there were dragonflies cruising the air with three-foot wingspans, and four-foot-long millipedes pushing their way through the undergrowth of the coal forests. Many paleontologists believe that these monsters could exist because the atmosphere was much richer in oxygen than nowadays. That alone would allow better respiration and larger size in the chitin-encased invertebrates.

Furthermore, it is easy to underestimate the capability of the insect brain. My favorite example is provided by the female of a fairyfly, one in a taxonomic group of extremely small parasitic wasps, which hatches from the egg of an underwater insect in which she has lived and grown up. She uses her legs as paddles to swim up to the surface. She digs through the tension of the surface film, and walks on top of it for a while. Then she flies in search of a mate, copulates, returns to the water, digs through the surface tension again, paddles to the bottom, searches until she finds an egg of the appropriate host insect, and lays one of her own inside it. The female fairyfly does all this with a brain almost invisible to the naked eye.

Equally impressive, honeybees and some species of ants can remember the location of up to five places where food is found and the time of day at each when food is available. Workers of an African hunting ant prowl singly over the forest far from their colony's nest. They circle and zigzag during the excursion. As they travel, they memorize the pattern of the foliage seen above their heads against the sky. Occasionally, they stop and look up to summarize where they are: upon catching an insect, they use this mental map to run home in a straight line.

How can an insect process so much information with a brain not much larger than the period below the question mark at the end of this sentence? The principal reason is the way the insect brain—much more efficient by unit volume—is constructed. Glial cells, which support and protect the brain cells of larger animals, including us, are omitted in the insects, allowing more brain cells to be packed into the same space. Also, each insect brain cell has many more connections on average to other cells than do those of vertebrates, allowing added communication by means of fewer information distribution centers.

So if I have rendered to your satisfaction at least plausibly the existence in a past era of high insect intelligence, let me go on to outline the morality and aesthetics of an imaginary termite-like civilization on another planet similar to our own, which I've based on Earth termites of the present day but bigger and raised to human-level intelligence. It's science fiction, of course, but unlike most such fiction, it is fully based on solid science.

SUPERTERMITE CIVILIZATION ON A DISTANT PLANET

Imagine, if you will, a vampire-like species that shuns the light of day, dying quickly if exposed to it. These termites come out to forage for food only if they must, and then only at night. They treasure complete darkness, high humidity, and constant heat. They eat rotting vegetable material. Some also consume fungi they grow in gardens mulched by rotting vegetation. As with some social insect species on Earth, only the king and queen are allowed to reproduce. The queen, her abdomen hugely swollen with ovaries, lies within the royal cell, doing almost nothing else but eat. She lays a constant stream of eggs, and occasionally mates with the little king who stands at her side. The hundreds or thousands of workers in the queendom, freed like human priests and nuns from sexual turmoil, devote their lives selflessly to rearing their brothers and sisters. A rare few of the young turn into virgin kings and queens, who leave the colony, find mates of their own, and start new colonies. The workers further attend to all of the other tasks, including education, science, and culture, of this supertermite civilization. Many of the inhabitants are soldiers, fitted with massive muscles and jaws and glands from which they spit poisonous saliva, ever ready for the chronic battles that break out among the colonies.

Life is spartan, and any deviation from the rules of the group, any attempt to reproduce or to attack others, is punished by death. Corpses of the workers that have died for any reason are eaten. Workers who grow ill or suffer injuries are also eaten. Communication is almost wholly by pheromones, from the tastes and scents of secretions released from glands located up and down the length of the body, as the source of our sound is in our larynx and mouth. Think of our human way in this remarkable line from Vladimir Nabakov's famous novel Lolita*: "Lo-lee-ta: the tip of the tongue taking a trip of three steps to tap, at three, on the teeth." Imagine then the release of pheromones from the line of pheromones in different combinations, different sequences, perhaps a trip of three stages in puffs of pheromone from the openings of glands along the side of the body. Pheromonal music, translated into sounds, might sound beautiful to us. It could unfold in melodies, cadenzas, beats, crescendos, and, with orchestras of the supertermites participating, symphonies, much more. All this would be experienced by smell.*

The supertermite culture would thus be radically different from ours, and extremely difficult to translate. The species would have its termitities as our species has its humanities. Yet—their science would be closely similar; its principles and mathematics could be mapped unambiguously onto our own. Supertermite technology might be more or less advanced, but it too would have evolved in parallel manner.

We would not like these supertermites, nor, I suspect, any other intelligent alien we encountered. And they would not like us. Each would find the other not just radically different in sense and brain, but morally repugnant. But this said, we could share our scientific knowledge to great mutual advantage. And, oh, before I forget to remind you. You don't need to engage in fantasy to envision cultures, or whole faunas and floras, on another planet. In fact, my extraterrestrial termites, minus culture, are based on the real mound-building termites of Africa.

Similar wonders await your attention. The universal nature of scientific knowledge yet to be revealed includes a near-infinitude of surprises.

New kinds of mussels and other novel organisms discovered in deep-sea hydrothermal vents on the Mid-Atlantic Ridge. Modified from original painting. © Abigail Lingford.

Sixteen

Searching for New Worlds on Earth

TO MAKE IMPORTANT DISCOVERIES anywhere in science, it is necessary not only to acquire a broad knowledge of the subject that interests you, but also the ability to spot blank spaces in that knowledge. Deep ignorance, when properly handled, is also superb opportunity. The right question is intellectually superior to finding the right answer. When conducting research, it is not uncommon to stumble upon an unexpected phenomenon, which then becomes the answer to a previously unasked question. To search for unasked questions, plus questions to put to already acquired but unsought answers, it is vital to give full play to the imagination. That is the way to create truly original science. Therefore, look especially for oddities, small deviations, and phenomena that seem trivial at first but on closer examination might prove important. Build scenarios in your head when scanning information available to you. Make use of puzzlement.

While I've spent a lot of time thus far on biology, obviously because I am a biologist, I am happy to emphasize that other fields of science yield comparable treasures of discovery. I've worked enough with mathematicians and chemists in particular to know that their heuristics—their process of making discoveries—is closely similar. Organic chemistry, for example, to substantial degree consists of exploring the almost endless array of possible molecules, and the occurrence of this chemodiversity in the natural world, and finally the physical and combinatorial properties of each kind of molecule. Take the elementary hydrocarbon CH_4 and run it in series up through C_2, C_3, C_4, and beyond, adding double and triple bonds, and sprinkle along the way the radicals S (sulfur), N (nitrogen), O (oxygen), and OH (hydroxy-), varying the form when possible into pure and branching strings, cycles, helices, and folds. The number of potential molecular "species" rises with molecular weight at a rate faster than exponential. Four million organic compounds were known by 2012, with 100,000 more being characterized each year, comparing favorably with 1.9 million biological species known and 18,000 new species added each year. Most of organic chemistry, and within it natural-products chemistry, consists of the study of the synthesis and characteristics of the molecules. Special attention is paid to those occurring in living organisms, where organic chemistry turns into biochemistry. Virtually all of life's processes and all of living structures are but the interplay of organic molecules. A cell is like a miniature rain forest, into which biochemists and molecular biologists conduct expeditions to find and describe organic structure, variety, and function.

The mind-set of astronomers is similar. They wander through the near-infinitude of space and time to find and describe the arrays of galaxies and star systems, and the forms of energy of matter within and between them. The development of particle physics has likewise been a journey into the unknown, to explore the ultimate components of matter and energy.

Across thirty-five powers of magnitude (powers of ten, hence of magnitude 1, 10, 100, 1000, and so on), from one subatomic particle to the entirety of the universe, science rules the enterprise of the human imagination applied to the laws of reality. Even if our intellect were somehow limited to the biosphere alone, scientific research would still be an endless adventure of exploration. Life invests the planet surface totally; no square meter is entirely free of it. Bacteria and microscopic fungi exist on the summit of Mount Everest. Insects and spiders are blown there by thermal drafts; and a few, including springtails and the jumping spiders preying on them, survive on the slopes close to the very top. At the extreme opposite in elevation, the bottom of the Mariana Trench in the western Pacific, thirty-six thousand feet below the ocean surface, bacteria and microscopic fungi flourish, and, with them, fish and a surprisingly large variety of single-celled foraminiferans.

There must be by definition somewhere on Earth a site with the greatest variety of organisms. The Yasuni National Park of Ecuador, which encloses a magnificent rain forest between the Rio Napo and Rio Curaray, is reputed to be that one biologically richest place on Earth. More precisely, its 9,820 square kilometers are believed to contain more species of plants and animals than any other piece of land of comparable area. The known roster supports the claim: recorded in the whole park are 596 bird species, 150 amphibian species (more than the number in all of North America), as many as 100,000 insects, and, growing in just a single average upland hectare, 655 tree species—also more than occur in all of North America. The only question about Yasuni's supremacy is whether there might exist some other, less explored section of the Amazon and Orinoco Basins that will prove even more diverse. At the very least, the Yasuni National Park is very close to the extreme of its kind. And outside the Amazon-Orinoco region, nothing in the world can approach it.

There is another reason to pay attention, not yet widely recognized even by most biologists: the Yasuni National Park may harbor the highest species numbers that have *ever* existed. Throughout the entire history of life, from the Paleozoic Era forward, 544 million years, the number of plant and animal species worldwide has been very slowly rising. Thus at the breakout from Africa and worldwide spread of *Homo sapiens*, beginning about 60,000 years before the present, Earth's biodiversity was likely at its all-time maximum. Then, extinction by extinction, human activity began to whittle the number down, and today that pace is accelerating. For the time being, Yasuni holds its own, and that is why it is recognized as a world treasure. We know only a fraction of the species of animals, especially the insects, found in the Yasuni, and next to nothing of their biology. We would like to take the full measure of this place, and of others of similar extreme high diversity, and come to understand the reason for its preeminence—before it is ruined by human greed.

In extreme opposition, there exists on Earth a

close outward approximation of the lifeless surface of Mars. In its own way it has been worth exploring. The place is the McMurdo Dry Valleys of Antarctica. To casual inspection the land seems as sterile as the surface of autoclaved glassware. But life is there, and it makes up the sparest and most stubborn of all of Earth's ecosystems outside the open surface of polar ice. Even though nitrogen is at the lowest concentration of any habitat on Earth, and water is almost nonexistent, it is surprising to find bacteria in the soil of the McMurdo Dry Valleys. The rocks scattered about seem lifeless, yet some are etched with almost invisible crevices in which communities of lichens live. These organisms are minute fungi that live symbiotically with green algae. They are concentrated layers just two millimeters beneath the surface of the rocks. Farther in, other such endoliths ("living in rocks") include bacteria capable of their own photosynthesis.

Scattered about in the McMurdo Dry Valleys are frozen streams and lakes, which contribute a small amount of moisture in the surrounding soil. The free water, which occurs in droplets and films, harbors small numbers of almost microscopic animals: tardigrades, the strange creatures sometimes called "bear animalcules" that I mentioned earlier, rotifers ("wheel animalcules"), and, most abundant of all, nematodes, also called roundworms. Although barely visible to the naked eye, the nematodes are the tigers of the land, the top of the food chain in this quasi-Martian world, and the antelope equivalents on which they feed are bacteria in the soil. In a few places can also be found rare mites and springtails, the latter a primitive form of insect. In all, sixty-seven species of insects have been recorded from the combined habitats of Antarctica, but only a few are free-living. The great majority are parasites that live in and on the warm plumage of birds and the fur of mammals.

As I write, there are many other places on the planet in which biological exploration has only begun. The greatest depths of the ocean, the abyss of eternal dark, consists of great submerged mountain ranges incised by deep unvisited valleys and intervening vast plains. The tips of many of the mountains rise above the water to form the oceanic islands and archipelagoes. Some come close but remain submerged. There are the seamounts. Their peaks are coated with marine organisms, many of whose species are unique to the location. The exact number of seamounts is still unknown. It has been estimated to run in the hundreds of thousands. Imagine the extent of human ignorance! Beneath the surface of the oceans and seas, which cover 70 percent of the Earth's surface, there exists an all but countless number of lost worlds. Their complete exploration will occupy generations of explorers from every discipline of science.

Life on Earth remains so little known that you can be a scientific explorer without leaving home. We have scarcely begun to map Earth's biodiversity at any level, from molecule to organism to niche in an ecosystem. Consider the following numbers of known

ORGANISM	NUMBER OF SPECIES ON EARTH KNOWN TO SCIENCE IN 2009	NUMBER OF SPECIES ESTIMATED TO BE ON EARTH IN 2009, KNOWN AND UNKNOWN
Plants	298,000	391,000
Fungi	99,000	1,500,000
Insects	1,000,000	5,000,000
Spiders & Other Arachnids	102,000	600,000
Mollusks	85,000	200,000
Nematodes (Roundworms)	25,000	500,000
Mammals	5,487	5,500
Birds	9,990	10,000
Amphibians (Frogs, etc.)	6,500	15,000
Fishes	31,000	40,000

and unknown species among different taxonomic groups of organisms around the world. They are why I like to call Earth a little-known planet. The data were pulled from global surveys made under the auspices of the Australian government in 2009.

The total number of species estimated in 2009 to have been discovered, described, and given a formal Latinized names worldwide was 1.9 million. The true number, both discovered and remaining to be discovered, could easily exceed 10 million. If the single-celled bacteria and archaea, the least known of all organisms, are added, the number might soar past 100 million. Five thousand kilograms of fertile soil contain, by one estimate, 3 million species, almost all unknown to science.

Why haven't scientists made more progress in exploring the world of bacteria and archaea? (The latter are an important group of single-celled organisms that outwardly resemble bacteria but possess very different DNA.) One reason for our ignorance is that a satisfactory definition of "species" in these organisms remains to be made. An even more important reason is that the different kinds of bacteria and archaea are so diverse in the environments they require in order to grow, and in the food they need to eat. Microbiologists have not learned how to culture the great majority of bacteria and archaea, in order to produce enough cells for scientific study. With the advent of rapid DNA sequencing, however, the genetic code of a strain can be determined with only a few cells. As a result, the exploration of species diversity has increased dramatically.

In citing these remarkable figures on biodiversity, I am not suggesting that you plan to become a taxonomist—although for now and many years ahead that would not be a bad choice. Rather, I wish to stress how little we know of life on this planet. When we also consider that the species is only one level in the hierarchy of biological organizations, molecule to ecosystem, then the immense potential of biology, and of all of the physics and chemistry relevant to biology, becomes immediately apparent.

If scientists know so little of raw biological diversity at the taxonomic level, we know even less of the life cycles, physiology, and niches of each species in turn. And for all but a very few localities on which biologists of diverse training have focused their energies, we are equally ignorant of how the idiosyncratic traits of individual species fit together to create ecosystems. Ponder these questions for a while: How do pond, mountaintop, desert, and rain forest ecosystems really work? What holds them together? Under what pressures do they sometimes disintegrate, and how and why? In fact, many are crumbling. Humanity's long-term survival depends on acquiring answers to these and many other related questions about our home planet. Time is growing short. We need a larger scientific effort, and many more scientists in all disciplines. Now I'll repeat what I've said when I began these letters: you are needed.

The female gypsy moth, located at the lower point of the active space, releases a pheromone cloud within which is a region of high concentration followed by the male. Drawing by Tom Prentiss (moths) and Dan Todd (active space of gyplure © *Scientific American*). Modified from "Pheromones," by Edward O. Wilson, *Scientific American* 208(5): 100–114 (May 1968).

Seventeen

The Making of Theories

THE BEST WAY I can explain the nature of scientific theories to you is not by abstract generalizations but by offering examples of the actual process of making theory. And because this part of science is the product of creative and idiosyncratic mental operations that are seldom put into words, I will stay as close to home as possible by using two such episodes in which I have been personally involved.

The first is the theory of chemical communication. The vast majority of plants, animals, and microorganisms communicate by chemicals, called pheromones, which are smelled or tasted. Among the few organisms that use sight and sound primarily are humans, birds, butterflies, and reef-dwelling fish. Working with the social behavior of ants in the 1950s, I became aware that these highly social insects use a variety of substances that are released from different parts of the body. The information they transmit is among the most complex and precise found in the animal kingdom.

As new information began to pour in, those of us conducting the early research saw that we needed a way to pull together the fragmented data and make sense of them. In short, we needed a general theory of chemical communication.

I was extremely fortunate during this early period to serve as the cosponsor of William H. Bossert, a brilliant mathematician working for a Ph.D. in theoretical biology. After completing his degree requirements in 1963, he was invited to join the Harvard faculty, and in a short time thereafter he received a tenured professorship in applied mathematics. While still a graduate student, he joined me in creating a theory of pheromone communication. The time was right for such an effort, and we were successful. On no other occasion in my scientific career has a project worked out as quickly and as well as did the collaboration with Bill Bossert.

To kick things off, I told him what I knew about the new subject. I laid out the basic properties of chemical communication as I had come to understand them. There was not a great deal of information to go on in this early period. From field and laboratory studies, I said, we knew that a wide variety of pheromones exist. It seemed logical that we should begin with a classification of the roles of all of those known, then try to make sense of each one in turn. The theory should deal not just with form and function of the pheromone molecules, which was the goal of most researchers, but also with their evolution. Put simply, we wanted to know what the pheromones are and how they work, of course, but also *why* they are one kind of molecule and not another.

Before giving you the theory, here are the specific "why" questions we meant for it to explain. Is the pheromone molecule used the best way possible, or is it one that was selected at random during evolution out of a limited array available for the job? What do the pheromone messages "look like" if you could see them spreading through space? Should the animal emit a lot of the pheromone or just a little in each message? How far and fast do the pheromone molecules travel through air or water, and why?

Here, then, in a nutshell, is the theory. *Each kind of pheromone message has been engineered by natural selection—that is, trial and error of mutations that occur over many generations resulting in the predominance of the best molecules, with the most efficient form of transmission allowed by the environment.* Suppose a population of ants is started by two ant colonies who compete with each other. The first colony makes one kind of molecule and dispenses it in a certain way, and the second colony makes another kind of molecule that is less efficient, or else is dispensed less efficiently, or both. The first colony will do better than the second, and as a consequence it will produce more daughter colonies. In the population of colonies as a whole, the descendants of the first colony will come to predominate. Evolution has occurred in the pheromone, or in the way it is used, or both.

Bossert and I agreed: "Let's think about ants and other organisms using pheromones as engineers." This thought took us quickly to ants recruiting other ants by laying a trail for them to follow. So, at the next picnic (or on your kitchen floor if the house is infested) drop a crumb of cake. It is logical to suppose that the ant scout that finds it needs to dribble out the trail pheromone at a slow rate in order to make the store of the substance she carries in her body last a long time. The piece of cake may be several ant-mile equivalents away. In this function, the ant is like an automobile engine designed for high mileage. In order to achieve such efficiency, the pheromone needs (in theory) to be a powerful odor for the ants following the trail. Only a few molecules should suffice. Also, the pheromone must be specific to the species using it, in order to provide privacy. It is bad for the colony if other ants from other species can pirate the trail, and even dangerous for the colony if a lizard or some other predator can follow the trail back to the nest. Finally, the trail substance should evaporate slowly. It needs to persist long enough for other members of the colony to track it to the end, and start laying trails of their own.

Then there are the alarm substances. When a worker ant or other social insect is attacked by an enemy, whether inside or outside the nest, it needs to be able to "shout" loud and clear, in order to get a fast response. The pheromone must therefore spread rapidly and continuously over a long distance. But it should also fade out quickly. Otherwise even small disturbances, if frequent, would result in constant pandemonium—like a fire alarm that cannot be turned off. At the same time, unlike the case for trail substances, there is no need for privacy. An enemy can gain little by approaching a location teeming with alert and aggressive worker ants.

Let me pause here to describe an easy way for you to smell an alarm pheromone yourself. Catch a honeybee from a flower in a handkerchief or other soft cloth. Squeeze the crumpled cloth gently. The bee will sting the cloth, and as it draws away it will leave the sting (which has reverse barbs) stuck in the cloth. When that happens, the immobile sting pulls out part of the bee's internal organs. Let the bee move to the side, then crush the sting and the organs between two fingers. You will smell an odor that resembles the essence of banana. Its source is a mixture of acetates and alcohols in a tiny gland located along the shaft of the sting. These substances function as an alarm signal, and they are the reason other bees rush to the same site and add their own stings. Next, if the eviscerated bee hasn't flown away, crush its head and smell that. The acrid odor you detect is from a second alarm substance, 2-heptanone, emitted by glands at the base of the mandibles. (Don't feel bad about killing a worker bee. Each has an adult life span of only about a month, and it is only one of tens of thousands that make up a colony. The colony in turn is potentially immortal, since new mother queens replace the old ones at regular intervals.)

The next category of pheromones are the attractants, in particular the sex pheromones, by which females call to males for the purpose of mating. The phenomenon is widespread not only in social insects but also throughout the animal kingdom. Other attractants also include the scent of flowering plants, in which the flowers call to butterflies, bees, and other pollinators. The most dramatic substances of the kind are the sex attractants of female moths, which can draw males upwind for distances of a kilometer or more.

Finally, Bossert and I reasoned in our initial classification, there are the identification substances. An ant, upon smelling these substances, can tell whether another ant is from the same or a different colony. It can also identify a soldier, ordinary worker queen, egg, pupa, or larva, and if the latter, its age. Carrying a chemical badge of this kind with you at all times means wearing the pheromone like a second skin. An identity pheromone is a single substance or, more likely, a mix of substances. It needs to evaporate very slowly and be detectable only at a very close range. If you closely watch one ant or some other social insect approach another, say while running along a trail or entering a nest, you will see the two sweep each other's body with their two antennae—a movement almost too fast for the eye to catch. They are checking body odor. If they detect the same odor, each will pass the other by. If the body odor is different, they will either fight or else flee from each other.

Reaching this point in the investigation, Bossert and I left the "adaptive engineering" method of evolutionary biology and passed into biophysics. We needed to envision the spread of the pheromone molecules from the body of the animal releasing them, and as precisely as possible. Obviously, as the pheromone cloud disperses, its density would decline—there would be fewer and fewer molecules in each cubic millimeter of space. Eventually there would be too few to smell or taste. Bossert then devised the crucial idea of "active space," within which the molecules are dense enough to be detected by the receiving plant, animal, or organism. He constructed models (at last, a place for pure mathematics!) to predict the shape of the active space. We were now in a new phase in creating the theory of pheromone communication.

With the ant or any other broadcasting organism sitting on the ground in still air, the shape of the active space would be hemispherical—one half of a sphere cut in two—with the broadcaster at the center of the flat surface. When an organism releases the pheromone from a leaf or object off the ground and in an air current, the shape of the active space would be an ellipsoid (roughly, shaped like an American football), tapering to a point at each end. The broadcaster would be at one of the points, releasing the pheromone downwind. When a trail is laid on the ground in quantities sufficient for it to be detected over a long period of time, the space would become a very long semiellipsoid, in other words an ellipsoid cut in half lengthwise at ground level.

Next we turned our attention of the design to the molecule itself. Trail substances and identification odors should consist, we reasoned, of either relatively large molecules or mixes of large molecules. They should diffuse slowly. Alarm pheromone molecules should be chosen in evolution to be smaller in size. They should form a more limited active space, and dissipate quickly. The qualities of the active space depend on five variables that can be measured: the diffusion rate of the substance, the surrounding air temperature, the velocity of the air current, the rate at which the pheromone is released, and the degree of sensitivity of the organism receiving it. With these measurable quantities in place, the theory began to take shape in a form that could be taken into the field and laboratory, and used to study animals as they communicated.

Next, we left biophysics for a while and entered the realm of natural products chemistry to learn the nature of the pheromone molecules. It's the same chemistry used widely in pharmaceutical and industrial research. It was our good luck that a recent major advance in molecular analysis put this part of the pheromone story within reach. By the late 1950s, the new technique of gas chromatography coupled with mass spectrometry made it possible to identify substances in quantities as little as a millionth of a gram, or less. Where previously chemists needed thousandths of a gram of pure substance to get the job done, now they needed only thousandths of a thousandth. The technique has allowed the detection of trace substances, including toxic pollutants, in the environment. Along with DNA sequencing (also requiring only a droplet of blood or the wipe of a wineglass), it also soon transformed forensic medicine. For us and other researchers it made possible the identification of pheromones carried in the body of a single insect. Ants commonly weigh between one and ten milligrams each. If a particular pheromone takes up only a thousandth or even a millionth of its body weight, it is still possible for researchers to make some progress in the characterization of the molecule. The chemists I worked with could obtain hundreds or thousands of ants. That was no great feat—it takes only a shovel and a bucket—and is one of the great advantages of working with ants. It became possible not only to isolate candidate pheromones but also to obtain enough of the material for bioassays—testing the material with live colonies to see if it evokes what theory suggests is the correct response.

In an early stage of pheromone research a biochemist, my friend John Law, and I set out to identify the trail substance used by the imported fire ant, which by that time had become one of the more noxious insect pests of the American South. We thought that in order to have plenty of the pheromone we should collect tens of thousands or even hundreds of thousands of the ants for extraction of the critical substance. That seemed quite practicable, because each fire ant colony contains upward of two hundred thousand workers. And I happened to know a way to gather that many fire ants quickly and efficiently. The imported fire ant, as a native to the floodplains of South America, has a unique way to avoid rising water. When the ants sense the approach of a flood from around and below them, they move to the surface of the nest, carrying with them all the young of the colony—the eggs, the grublike larvae, and the pupae—while nudging the mother queen upward as well. When the water reaches the nest chambers, the workers form a raft of their bodies. The whole colonial mass then floats safely downstream. When the ants contact dry land, they dissolve their living ark and dig a new nest.

It occurred to me that if we simply excavated fire ant nests and dumped them and the soil into nearby pools of water, the colony would rise to the surface and gather as an ant-pure raft while the dirt settled to the bottom. We tried this crude method on roadsides outside Jacksonville, Florida, and it worked. We came back with the requisite one hundred thousand worker ants (roughly estimated, not counted!) and my hands covered with itching welts from the stings of many angry ants.

Back at Harvard in Law's laboratory, the search for the fire ant trail pheromone at first went well. The crucial substance appeared to be a relatively simple molecule—a terpenoid—and its complete molecular structure seemed within reach. Then came frustration, and a mystery. As the chemists attempted to purify the substance in order to characterize it definitively, and we proceeded to assay the reactions they produced by laying artificial trails in the laboratory, the response to the fraction supposedly containing the pheromone grew progressively weaker. Was the pheromone an unstable compound? Thinking that to be a good possibility, and concluding that the substance probably couldn't be identified with the equipment and material available, we quit. To help others making the attempt we published a note in the science journal *Nature*—one of the few articles the editors have ever accepted that reported a failed experiment.

Years later, Robert K. Vander Meer, a natural product chemist working on fire ant pheromones in Florida, discovered the reason for our failure. The trail substance, it turned out, is not a single pheromone, but a medley of pheromones, all released from the sting onto the ground. One attracts nestmates of the trail layer, another excites them into activity, and still another guides them through the active space created by the evaporating chemical streaks. All of the components need to be present to evoke the full response in a fire ant worker seen in the field and laboratory. By not realizing this complexity, and thereby taking aim only at one of the components, we had failed to identify any of them.

In the 1960s and 1970s research on pheromones deepened and expanded, becoming an important part of the new discipline of chemical ecology. Researchers worked out with increasing accuracy what proved to be the complex pheromone codes of ant and honeybee colonies. Our theory of engineering by natural selection proved out well. However, recognizing that we had dealt with biology, and the independent events of natural selection, the correlations we proposed were only roughly met. A few strange, idiosyncratic exceptions were found, some of which to this day await further theory and experimental testing.

Ecosystems, with their rich complexes of interacting plants, animals, fungi, and microorganisms, came to be seen in a new way and the theories that guide ecology were altered accordingly. There was a different sensory world to be understood, one wholly invisible to human sight and hearing. The signals are in the air, spread over the ground, and beneath in the soil, and in pools of water. They form a crisscrossing of odors and scents, a riot of voices unheard by us that variously broadcast, threaten, or summon: *Check me as I approach you, I am a member of your colony. I have discovered an enemy scout, now hurry, follow me. I am a plant whose flowers have opened up this night and I wait here for you, come to me for a meal of pollen and nectar. I am a female cecropia moth calling, so if you are a male cecropia moth, follow my scent upwind, come to me. I am a male jaguar, alone on my territory, if you have detected this scent, you are trespassing, so get out, get out now.*

By science and technology we have entered this world, but we have only begun to explore it. Only when it becomes better known will we gain a part of the knowledge needed to understand how ecosystems are put together and, from that, how to save them.

Now I hope you see how theories are made, and how they work. The process can be messy, but the product can be true and beautiful. As factual information grows about any subject—in this case chemical communication—we dream about what it all means. We make propositions about how the phenomena we discovered work and how they came into existence. We find a way to test these various hypotheses. We look for a pattern that emerges when we put the parts together, like a jigsaw puzzle. If we find such a pattern, it becomes the working theory—we use it to think up new kinds of investigation, in order to move the whole subject forward. If this extension doesn't work very well and now facts appear that contradict the theory, we adjust it. When things get bad enough, we junk the theory and create a new one. With each such step, science moves closer to the truth—sometimes rapidly, sometimes slowly. But always closer.

Woolly mammoths, a now-extinct species of the World Continent Fauna. Modified from the original painting. © Natural History Museum Picture Library, London.

Eighteen

Biological Theory on a Grand Scale

My second example of the growth of theory is from biogeography, the science that explains the distribution of plants and animals. In its global reach of space and time, biogeography is the ultimate discipline of biology—in the same sense that astronomy is the ultimate discipline of the physical sciences. When the mapping of species around the world is added to the study of how they got there, biogeography acquires a noble grandeur. At least, that was how I felt when as a college student in my late teens I looked up from my studies of descriptive natural history to study the processes of evolution. I learned to ask: What kind of process creates biodiversity? What other kind scatters species into their current geographic ranges? Neither kind occurs at random, I read. Both are the products of understandable causes and effects. I was already totally devoted to making a career in natural history, as an expert on insects. A government entomologist, perhaps, or a park ranger, or a teacher. Now I rejoiced. I could also be a real scientist!

The first revelation for me came from the Modern Synthesis of evolutionary theory. Put together mostly in the 1930s and 1940s, it united the original Darwinian theory of evolution by natural selection with advances being made in the modern disciplines of genetics, taxonomy, cytology, paleontology, and ecology. I was especially impressed by Ernst Mayr's 1942 synthesis, *Systematics and the Origin of Species*, which I could immediately apply to my knowledge of taxonomy, the systematic classification of organisms. Suppose you yourself were working on a particular subject, say the colors of gems or the taste of wines, and you came upon a theoretical work that seemed to make sense of everything you already knew. You would have the same kind of transformative experience.

Later, as a graduate student at Harvard, I discovered a remarkable work on theory of biogeography only occasionally noticed by previous scientists: William Diller Matthew's "Climate and Evolution," published in a 1915 issue of *Annals of the New York Academy of Sciences*. In it the eminent vertebrate paleontologist, who worked as a curator of mammals in the American Museum of Natural History located in New York City, proposed a grand scheme for the origin and spread of mammals around the world. The kinds of mammals destined to be dominant in this way have originated, he wrote, in the great Eurasian landmass of the north temperate zone, roughly present-day England all the way to present-day Japan. Being competitively superior, they eliminated older, formerly dominant groups that had occupied the same niches. The early rulers were not extinguished entirely, however. They still flourished in areas not yet colonized by the newcomers. Think of the present great northern landmass formed by Europe, northern Asia, and North America as the hub of a wheel. To the south, Matthew said, tropical Asia to Africa, Australia, and Central and South America are the spokes of the wheel. Dominants originate in the hub and spread through the spokes. At the time of his account, Matthew's theory seemed to fit the facts.

The dominant groups of the North, Matthew went on, are superior because they evolved in rugged, severely seasonal climates, which required a general toughness and ability to adapt to change. These most recent winners include animals familiar to all Eurasians and North Americans: mice and rats (taxonomic family Muridae), deer (Cervidae), cattle (Bovidae), weasels (Mustelidae), and, of course, us (Hominidae). Former dominants, now confined to the southern spokes, are the rhinoceroses (Rhinoceratidae), elephants (Elephantidae), and primates exclusive of man.

Right or wrong, and by evidence available in Matthew's time it seemed right (although much less so now), I saw the theory as prehistory on a global scale. It was biology lifted to the maximum in space and time. And it was scientific natural history, the subject I had chosen!

In 1948 Philip J. Darlington, whom years later I was to succeed as curator of insects at Harvard's Museum of Comparative Zoology, presented a different story for the reptiles, amphibians, and freshwater fishes, no less grand than that of Matthew's for mammals. These cold-blooded vertebrates, he said, arose not in the north temperate zone as supposed by Matthew for the warm-blooded mammals, but in the vast tropical forests and grasslands that once covered most of Europe, northern Africa, and Asia. They then spread south into the peripheral continents, much reduced in diversity of species, and northward into the north temperate zone. It also turned out from the new wave of fossil research that humanity originated not in Eurasia but in the tropical savannas of Africa.

I was raised, so to speak, more on Darlington than on Matthew, but Matthew I found to be right in one important respect. There was indeed a global pattern of dominant groups arising in large, ecologically varied portions of the world's landmass.

Then came the equally grand theory of the World Continent Fauna, the existence of which supported the overall theme developed by both Matthew and Darlington. For tens of millions of years South America was isolated from North America by a broad seaway that submerged the present-day Isthmus of Panama, thereby connecting the Pacific Ocean to the Caribbean Sea and isolating the continents on either side. Mammals, except bats, as a rule could not cross the broad stretch of ocean water. As a result those in South America evolved independently from those in North America. But the two faunas converged in outward appearance and in the niches they filled. In the north there were horses, in the south horselike litopterns. The rhinos and hippos of the north were duplicated, roughly, by South American toxodonts, and tapirs and northern elephants respectively by southern astrapotheres and pyrotheres. Shrews, weasels, cats, and dogs were matched in varying degree by the diverse members of the South American family Borhyaenidae. The fearsome saber-toothed tiger of North America was approached in overall appearance by an equivalent in South America, even though they remained very different in another way: the northern saber-tooth was a placental (fetus carried throughout in the uterus) and the South American one was a marsupial (fetus carried part of the way in an outside pouch).

This evolutionary convergence was the greatest on the land that the world has ever seen. Imagine that we could travel back in time to South America as it was ten million years ago and make a safari across its savanna, much as tourists do today in East Africa:

> *Say we are there back then on the edge of a lake, early one sunny morning, turning our gaze slowly through a full circle. The vegetation looks much like modern savanna. Out in the water a crash of rhinoceros-like animals browse belly-deep through a bed of aquatic plants. On the shore something resembling a large weasel drags an odd-looking mouse into a clump of shrubs and disappears into a hole. A creature vaguely like a tapir watches immobile from the shadows of a nearby copse. Out of the high grass a big, catlike animal suddenly charges a herd of—what?—animals that are not quite horses. Its mouth is thrown open nearly 180 degrees, knife-shaped canines projecting forward. The horse look-alikes panic and scatter in all directions. One stumbles, and . . .*

This independent kingdom of wildlife disappeared over a million years ago, long before the arrival of human beings, while its North American equivalent persisted mostly intact until only about ten thousand years ago, after skilled human hunters arrived and began to spread over the continent. Each appeared to have reached a balance within its own domain. Why, then, did the southern kingdom decline while the northern kingdom lived on?

This obvious disparity in survival brought biogeographers to the interesting question implied by the balance of nature: What happens when two full-blown, closely similar dynasties meet head-on? If it were possible to play God with geological spans of time to wait and watch, the ideal experiment would be this: Allow two isolated parts of the world to fill up with independent adaptive radiations of plants and animals, so that the majority of species in each theater have close ecological equivalents in the other theater; then connect the two regions with a bridge and see what happens. When the organisms intermingle, would those from one theater replace the other, so that a single fauna and flora comes to occupy the entire range?

The grand experiment has in fact been performed once in relatively recent geological time, and we can deduce a great deal of what happened by comparing fossil and living species. Two and a half million years ago the Isthmus of Panama rose above the sea, bridging the ancient Pacific-to-Caribbean seaway and allowing the mammals of South America to mingle with the mammals of North and Central America. Species from each continent spread into the other.

The change in biodiversity that occurred can best be measured at the taxonomic level of the family. Examples of mammalian families are the Felidae, or cats; Canidae, dogs and their relatives; Muridae, the common mice and rats; and of course Hominidae, human beings. The number of mammalian families in South America before the interchange was thirty-two. It rose to thirty-nine soon after the Isthmus of Panama connection, and then subsided gradually to the present-day level of thirty-five. The history of the North American fauna was closely comparable: about thirty families before the interchange, rising to thirty-five, and subsiding to thirty-three. The number of families crossing over was about the same from both sides.

When all this information was put together, the stage was set for another kind of theory. When biologists see a number go up following a disturbance and then fall back to the original level, whether body temperature, density of bacteria in a flask, or biological diversity on a continent, they suspect that an equilibrium exists in the system. The restoration of the numbers of mammalian families in both North and South America points to such a balance of nature. In other words, there appears to be a limit to diversity, in the sense that two very similar major groups cannot coexist in their fully radiated condition. A closer examination of the ecological equivalents on both continents, dwellers in the same broad niche, reinforces this conclusion. In South America marsupial big cats and smaller marsupial predators were replaced by their placental equivalents. Toxodonts gave way to tapirs and deer. Still some unusual specialists—the wild cards—were able to persist. Anteaters, tree sloths, and monkeys continue today to flourish in South America, while armadillos are not only abundant throughout tropical America but are represented by one species that has expanded its range throughout the southern United States.

In general, where close ecological equivalents met during the interchange, the North American elements prevailed. In this part of the world at least, Matthew's theory was vindicated. The North American mammals also attained a higher degree of diversification, as measured by the number of genera. A genus is a group of related species and a group of genera is a family. The genus *Canis*, for example, comprises domestic dogs, wolves, and coyotes; other genera in the dog family Canidae include *Vulpes* (foxes), *Lycaon* (African wild dogs), and *Speothos* (South American bush dogs). During the interchange, the number of genera rose sharply in both North and South America and remained high thereafter. In South America it began at about seventy and has reached 170 at the present time. The swelling of numbers has come principally from speciation and radiation of the World Continent mammals after they arrived in South America. The old, pre-invasion South American elements were not able to diversify significantly in either North or South America. So the mammals of the Western Hemisphere as a whole now have a strong northern cast. Nearly half of the families and genera of South America belong to stocks that have immigrated from North America during the past 2.5 million years.

Why did the northern mammals prevail? No one knows for sure. The answer has been largely concealed by complex events imperfectly preserved in the fossil record—the paleontologist's equivalent of the fog of war. The question remains before us, part of the larger unsolved problem toward which our understanding of dynastic succession is directed. Evolutionary biologists keep coming back to it compulsively, as I did one night while camping at Fazenda Dimona, in the Brazilian Amazon, surrounded by mammals of World Continent origin. What comprises success and dominance?

Success in biology is an evolutionary idea. It is best defined as the longevity of a species with all its descendants. The longevity of the Hawaiian honeycreepers will eventually be measured from the time the ancestral finchlike species split off from other species, through its dispersal to Hawaii, and finally to that time when the last honeycreeper species ceases to exist.

Dominance, in contrast, is both an ecological and evolutionary concept. It is best measured by the relative abundance of the species group in comparison with other, related groups, and by the relative impact it has on the life around it. In general, dominant groups are likely to enjoy greater longevity. Their populations, simply by being larger, are less prone to sink all the way to extinction in any given locality. With greater numbers, they are also better able to colonize more localities, increasing the number of populations and making it less likely that every population will suffer extinction at the same time. Dominant groups often are able to preempt the colonization of potential competitors, reducing still further the risk of extinction.

Because dominant groups spread farther across the land and sea, their populations tend to divide into multiple species that adopt different ways of life: dominant groups are prone to experience adaptive radiations. Conversely, dominant groups that have diversified to this degree, such as the Hawaiian honeycreepers and placental mammals, are on average better off than those composed of only a single species: as a purely incidental effect, highly diversified groups have better balanced investments and will probably persist longer into the future. If one species comes to an end, another occupying a different niche is likely to carry on.

The mammals of North American origin proved dominant as a whole over the South American mammals, and in the end they remained the more diverse. Over two million years into the interchange, their dynasty prevails. To explain this imbalance, paleontologists have forged a widely held theory, an evolutionary-biologist kind of theory, in other words a rough consensus consistent with the largest number of facts. The fauna of North America, they note, was not insular and sharply different like that of South America. It was and remains part of the World Continent Fauna, which extends beyond the New World to Asia, Europe, and even Africa. The World Continent is by far the larger of the two landmasses. It has tested more evolutionary lines, built tougher competitors, and perfected more defenses against predators and disease. This advantage has allowed its species to win by confrontation. They have also won by insinuation, like raccoons and pack-forming wild dogs; many were able to penetrate sparsely occupied niches more decisively, radiating and filling them quickly. With both confrontation and insinuation, the World Continent mammals gained the edge.

The testing of this theory, first conceived on a rough grand scale by William Diller Matthew and Philip Darlington, has just begun. Right or wrong, whether decisive in empirical support or not, its pursuit alone promises to link paleontology in interesting new ways to ecology and genetics. That synthesis will continue as the study of biological diversity expands in widening circles of inquiry to other disciplines, to other levels of biological organization, and across farther reaches of time. You have a place in it if animals and plants interest you in their own right, and especially if you like epics and the clash of worlds.

The author identifying insects at an osprey nest, Florida Keys, March 19, 1968. Photograph by Daniel Simberloff.

Nineteen

THEORY IN THE REAL WORLD

IT MAY SEEM to you that science, having grown so large and complex in fact and theory, would be a difficult profession to enter. Perhaps you worry that most of the opportunities in research and application are closed, that competition for the rest is tight and daunting, and most of the epics and big pictures have been filled in. You would be wrong. The researchers of my generation and others before you accomplished a lot. But they did not close all pathways and enter all unknown regions. Instead, they opened new ones. In science every answer raises more questions. I will ramp up that important truth to an exponential degree: in science every answer creates *many* more questions. Thus has it ever been, even before Newton held up a prism to a sunbeam and Darwin puzzled over variation among the Galápagos mockingbirds.

It was also Newton who famously said, for all

scientists into the future, "If I see further than others, it is by standing on the shoulders of giants." I will now tell you a story of shoulders and giants.

It could begin at any one of several times, but I will start on December 26, 1959, at the annual meeting of the American Association for the Advancement of Science, in Washington, D.C., when a mutual friend introduced me to Robert H. MacArthur. Robert (he resisted being called Bob) and I were relatively young. He was twenty-nine and I was thirty. We were both very ambitious, each searching self-consciously for the opportunity to make a major advance in science. MacArthur was brilliant. He was widely thought the new avatar of theoretical ecology, having already made several seminal advances. He was an avid naturalist and expert on birds, and in addition (very important in our case) an able mathematician. Thin, sharp in face and disposition, he had an intense and withdrawn manner that warned off fools. He was not the kind who placed hand on shoulder and slapped backs, nor did he often laugh out loud. Although we spent a great deal of time together, MacArthur and I never became close friends. Looking back today, I realize we never finished taking the measure of each other.

His mentor at Yale, the first giant in this story, had been G. Evelyn Hutchinson, who was bringing ecology into the Modern Synthesis of evolutionary biology. He was famous for the earnest brilliance of his students. Under his tutelage, MacArthur had already made his mark by showing how complex ecological processes such as competition in community organization and the evolution of reproductive rates could be simplified into a form amenable to useful mathematical analysis. We were both, ten years later, to be elected to the U.S. National Academy of Sciences, also at an exceptionally young age. In 1972, at the peak of his creativity, MacArthur died of kidney cancer. Science was thus stripped of his future greatness, a huge loss.

Coming together for meetings during the early 1960s, we both saw ecology and evolutionary biology as potentially one continuous discipline filled with opportunity for innovation in theory and field research. This was a new concept heralded by G. Evelyn Hutchinson. But we had another, equally pressing motivation. By the 1960s, the revolution of molecular and cellular biology was already well under way. The second half of the twentieth century was clearly going to be their golden years, and one of the most transformative periods of all time in the history of science. Molecular biology and cellular biology were propelled not only by the extraordinary opportunities they provided for innovation, but also by the massive funding they received due to their obvious relevance to medicine.

MacArthur and I understood clearly what was happening. We also saw that one negative result in science was the proportionate downgrading of our own disciplines, ecology and evolutionary biology. We had no equivalent of the double helix, no direct link to physics and chemistry, as did molecular and cellular biology. Rachel Carson's seminal *Silent Spring* had been published in 1962, launching the modern environmental movement, which might have provided a nourishing source of funding equivalent to medicine, but that beneficence was still in its infancy. The new disciplines of conservation biology and biodiversity studies did not emerge until the 1980s.

Furthermore, aside from population genetics and some very abstract principles of ecology, we had few ideas that could be solidly linked together in the expected manner of mature natural sciences. Molecular biologists and cellular biologists were filling faculty openings in research universities, unconcerned about biology at the levels of the organism and of the population. In their judgment, if they bothered to form one at all, our disciplines were old-fashioned and hopelessly unproductive. The frontiers of biology, it appeared, had shifted decisively leftward, in the direction of physics and chemistry. It was not so much that this new generation of biologists considered the old guard unimportant. It was more that they expected to do a better job of the research when, someday, they got around to it themselves. The pathways were there for MacArthur and me and other young ecologists, but they proved difficult to follow.

My difficulties at Harvard were intensified by the fact that I was the only young tenured Harvard professor in what was later to be called organismic and evolutionary biology. The elder and more distinguished faculty members in the same disciplines were either wholly absorbed in tending their personal academic gardens or else in denial—aloof and disinclined to deal with the threat.

The ultimate in *noblesse non oblige* was the venerable George Gaylord Simpson, the second giant in the story. He was a world authority in vertebrate paleontology and one of the authors of the Modern Synthesis. He had devised a brilliant picture of the evolution and movement of faunas around the world. But his withdrawal from engagement with others was legendary. Aging and ill by the time he came to Harvard, crippled by a falling tree during a recent visit to the Amazon, he preferred to work alone in his office deep in the bowels of the Museum of Comparative Zoology. When on one occasion Robert MacArthur visited the Department of Biology, I made an appointment for him to see Simpson. A meeting of first-rate minds, I thought, across the generations. I escorted him to the great man's office, then left the two alone so as not to intrude on their conversation.

(I expected to hear all about it later anyway.) I returned to my office and began to catch up on some paperwork. Scarcely fifteen minutes later MacArthur reappeared at my door. "He hardly said a word," Robert reported. "He just refused to talk."

Simpson's taciturnity, and from my viewpoint his indifference toward addressing the intellectual imbalance of biology at Harvard, had already played a role in the introduction of the term "evolutionary biology." In 1960, the faculty members of the Department of Biology working on ecology and evolution, being outgunned and outfunded and soon to be outnumbered, decided to form a committee to organize and unify our efforts. I arrived early for the first meeting, and soon was followed by Simpson, who sat across from me (silently) to await our colleagues.

"What shall we call our subject?" I ventured.

"I have no idea," he responded.

"What about 'real biology'?" I continued, trying for humor. Silence.

"Whole-organism biology?"

No response. Well, those were bad ideas anyway.

There was a pause, then I added, "What do you think of 'evolutionary biology'?"

"Sounds all right to me," Simpson said, perhaps just to keep me quiet.

Other committee members began to file in, and when all were settled, I seized the opportunity to assert, "George Simpson and I agree that the right term for the overall subject we represent is 'evolutionary biology,'" the name I had made up on the spot.

Simpson said nothing, whereupon our group became the Committee on Evolutionary Biology. In time it grew to be the official Department of Organismic and Evolutionary Biology. Thus was born the name of a scientific discipline. If there was an earlier independent birth elsewhere, and I've heard of none, at least the most influential use of the name was made at a time when it was most needed.

Envy and insecurity are among the drivers of scientific innovation. It won't hurt if you have a dose of them also. For MacArthur and myself, the desire to create a new theory was reinforced by the recognition that what we were now calling evolutionary biology, and its more quantitative subdivision of population biology, required a rigor comparable to that of molecular and cellular biology. We needed quantitative theory and definitive tests of the ideas spun from the theory and vivid connections to real-life phenomena. Such hallmarks of excellence were relatively sparse in the subjects our efforts addressed. It was time to search for them in a focused manner.

I spoke to MacArthur about islands I had visited around the world, and their use in studying the links between the formation and geography of species. I could see that he was not thrilled by the complexity of the subject. He became much more interested in the species-area curves that I had also been plotting. These displayed in a simple form the geographic areas (as in square miles or square kilometers) of islands in different archipelagoes of the world, principally the West Indies and western Pacific, and the number of bird, plant, reptile, amphibian, or ant species found on each island. We could see plainly that with an increase of area from one island to the next, the number of species increased roughly to the fourth root. This means, for example, that if one island in an archipelago is ten times the size of another in the same archipelago, it would contain approximately twice the number of species. We also observed that islands more distant from the mainland had fewer species than those close by.

When I talked about equilibrium I spoke of the islands near and far as being "saturated." MacArthur said, "Let me think about this for a while." I trusted him to come up with something. I'd already seen evidence of MacArthur's ingenuity in breaking down complex systems into simpler ones.

MacArthur soon wrote a letter to me in which he postulated the following:

> *Start with an empty island. As it fills up with species there are fewer species available from other islands to*

arrive as immigrants, and so the rate of immigration falls. Also, as the island fills up with species, it becomes more crowded and the average population size of each species decreases. As a result the rate of species extinction rises. Therefore, as the island fills up, the immigration rate falls, and the extinction of the species already present rises. Where the two curves cross, the extinction rate equals the immigration rate, and the number of species is at equilibrium.

To continue, on small islands the crowding of the species is more severe, and the extinction rate curve is steeper. On distant islands, immigration is less, and the immigration curve less steep. In both cases the result is a smaller number of species at equilibrium.

In 1967, MacArthur and I applied this simple model with every scrap of data on related subjects in ecology, population genetics, and even wildlife management we could find, and fitted it together, as best we could, in *The Theory of Island Biogeography*. The book enjoyed and continues to enjoy considerable influence in the disciplines from which it was constructed. It also played a role in the creation of the new discipline of conservation biology during the decades to follow. It was a good example of the principle I've urged you to remember: in research define a problem as precisely as possible, and choose if need be the one or two partners needed to solve it.

Even so, I wasn't completely satisfied with our product. I asked myself even while it was unfolding, how can we put such theory to a test? The equilibrium we envisioned might require centuries to achieve. So, how does one conduct an experiment with Cuba, Puerto Rico, and the other islands of the West Indies? One doesn't. Instead one looks for another, more tractable system. You may recall another principle of scientific research I offered you in an earlier letter. It is that for every problem there exists a system ideally suited for its solution. In biology the system is usually an organism of a particular species, such as the bacterium *Escherichia coli* for problems in molecular genetics. I was looking for something located higher in the scale of biological organization. I needed an ideal ecosystem.

I was driven by two intense desires. I wanted to go on working on islands, whatever the excuse. And I wanted to do something radically new in biogeography. I reasoned that I might accomplish both if I chose an ecosystem small enough to be manipulated.

A solution then presented itself. Insects—my specialty—are almost microscopic in size compared to the mammals, birds, and other vertebrates that had been featured in earlier biogeographic studies. They weigh a few milligrams or less, where vertebrates are measured in grams or more. There are large numbers of tiny islands on which insects can live and breed for generations. Instead of just one or several islands the size of Cuba, Barbados, or Dominica, where birds and mammals can be studied, there are hundreds of thousands of islands around the world with an area of a hectare or less. Somehow, I thought, the insect, spider, and other invertebrate faunas of a few could be altered so that the rates of immigration and extinction onto them could be measured. From these data multiple tests could be devised to test hypotheses, to evaluate theory itself, and to discover new phenomena.

A new world opened in my imagination. I saw the islets of the world as the perfect model ecosystem. Now I sought a laboratory. It had to be a cluster of little islands, variously big and small, close and distant. Where might such an ideal micro-archipelago be? I scanned detailed maps of the eastern Atlantic and southern Gulf Coast of the United States, from the rocky prominences of Maine and the Harbor Islands of Boston to the barrier islands of the Carolinas, Georgia, Florida, and the Gulf states to the west. All could be reached in a day's travel from Harvard University. It did not take long to settle on the multitudinous tropical islands of the Florida Keys and Florida Bay.

To conduct experiments that would yield what scientists like to call "robust" conclusions, I needed to have my islets start from zero—empty, harboring no insects at all. My attention fixed on the small, wave-washed islands of the Dry Tortugas, the outermost cluster of the Florida Keys. Except for Fort Jefferson at the very end, they are almost all desert islands, harboring only small patches of vegetation and relatively few species of insects and other invertebrates in residence. There was an advantage to their simplicity: whenever a hurricane crosses over them, they are swept clean of terrestrial life.

In 1965 I took a team of graduate students with me to the Dry Tortugas to look over the situation. We mapped every plant on several of the islands and recorded every insect and other invertebrate species we could find. During the next hurricane season, in 1966, not one but two hurricanes crossed the Dry Tortugas. We returned soon thereafter, and sure enough, the small islands were bare of plants and terrestrial animals.

It seemed that the main problem had been solved, but by this time I had begun to have doubts about using the Dry Tortugas. I believed that in order to conduct a high-quality experiment of lasting value, the kind that others could replicate conveniently, I needed a better laboratory. I wanted more islands than those that make up the Dry Tortugas. I needed to conduct the removal of the species myself, and not rely on random weather. It would also be best to use controls—islands closely identical to the experimental set, and treated the same but without removing the animals. Finally, I needed more biology. The faunas of the Dry Tortugas are so small and the life spans of the ecosystems so short as to reduce their faunas and floras to random number generators. I needed larger faunas more typical of natural ecosystems, and I needed less disturbed islands.

Before telling you how the goal was accomplished, I will pause to reinforce a point I made earlier: that successful research doesn't depend on mathematical skill, or even the deep understanding of theory. It depends to a large degree on choosing an important problem and finding a way to solve it, even if imperfectly at first. Very often ambition and entrepreneurial drive, in combination, beat brilliance.

I was determined to solve this problem of biogeography, and was excited by the challenge of developing a new technology doing it. I found what I needed in the small mangrove islands of Florida Bay, just to the north of the Dry Tortugas. There are a lot of them: consider the implication of the archipelago at the northern end of the bay called the Ten Thousand Islands. The damage done to the entire Florida Bay mangrove system by removing the invertebrates from a dozen or so would be negligible, and soon repaired.

At this point I enlisted the collaboration of Daniel S. Simberloff, one of my graduate students with a strong background in mathematics. I quickly realized I had chosen a partner wisely. As with MacArthur's work, Simberloff's mathematics fitted my own natural history nicely. From this point forward, while facing the unknown together, we became more colleagues than teacher and student. Together, step by step, we worked out the method of removing all of the invertebrate animals from the mangrove islets without damaging the trees and other vegetation. Without detailing our failures and false starts to you here, we devised the simple and straightforward method of eradication: hire a pest control company to erect a tent over each island and fumigate it. That was not as easy as it sounds. Working as a team, we had to invent the right framework to be erected in shallow water and find the right kind and dosage of dispersible insecticide to use. We had to walk through gluelike muck, and convince the workers helping us that the ground sharks swimming in close to the islands at high tide were harmless.

Not least, Simberloff and I also had to create a network of experts on the various groups of invertebrates—beetles, flies, moths, barklice, spiders, centipedes, and so on—in order to identify the species correctly.

After two years of monitoring the immigrations and extinctions that followed, and to my great relief (and Simberloff's also—he had to get a Ph.D. thesis out of his part of the work), the recolonization fitted the equilibrium model. We also learned a great deal about the colonization process itself. I found the whole of the adventure, from theory to experiment, one of the most satisfying experiences of my entire scientific life.

I hope that in your own career you will see one or more opportunities of this kind and, like Daniel Simberloff and myself, find the risk worth taking. We stood on the shoulders of giants and were able to see a little bit farther.

V

TRUTH *and* ETHICS

The U.S. National Medal of Science.

Twenty

THE SCIENTIFIC ETHIC

I HAVE COME TO the end of my counsel to you, and will now close these letters with advice on proper behavior in the conduct of your research and publication.

You are not likely to be directly pressed during your career on such largely philosophical questions as the propriety of creating artificial organisms or conducting surgical experiments on chimpanzees. Instead, by far the greatest proportion of moral decisions you will be required to make is in your relationships with other scientists. Entrepreneurial endeavor beyond the level of puttering creates difficulties other than the mere risk of failure. It will put you into a competitive arena for which you may not be emotionally prepared. You may find yourself in a race with others who have chosen the same track. You will worry that someone better equipped and financed will reach the goal before you. When multiple investigators create an important new field simultaneously, they often create a golden period of excited cooperation, but in later stages, as different groups follow up on the same discoveries, some amount of rivalry and jealousy is inevitable. For you, if successful, there will be gentle competitors and ruthless competitors. There will be gossip and some protective secrecy. That should come as no surprise. Business entrepreneurs suffer when competitors beat them to the marketplace. Should we expect scientists to be different?

Original discoveries, to remind you, are what count the most. Let me put that more strongly: they are *all* that counts. They are the silver and gold of science. Proper credit for them is therefore not only a moral imperative, but vital for the free exchange of information and amity within the scientific community as a whole. Researchers rightly demand recognition for all their original work, if not from the general public then from colleagues in their chosen field. I have never met another scientist who was not pleased—deeply pleased—by a promotion or award bestowed for original research. As the actor Jimmy Cagney said of his career in motion pictures, "You're only as good as people say you are."

The great scientist who works for himself in a hidden laboratory does not exist. Therefore, be rigorous in reading and citing literature. Bestow credit where it is deserved, and expect the same from others. Honest credit carefully given matters enormously. Recommending a colleague for awards or other forms of recognition is a relatively uncommon form of altruism when practiced among scientists. Even if it proves difficult, do not shrink from taking that step. On the other hand, granting it to a rival, especially one you do not like and at the risk of your own recognition, would be true nobility. It is not expected of you. Let others make the nomination. Instead just grit your teeth and extend your congratulations.

You will make mistakes. Try not to make big ones. Whatever the case, admit them and move on. A simple error in reporting or conclusion will be forgiven if publicly corrected. (At least one leading journal has a special erratum section.) An outright retraction of a result will not cause permanent harm if done graciously, and especially with thanks to the scientist who reported the error with evidence and logical reasoning. But never, ever will fraud be forgiven. The penalty is professional death: exile, never again to be trusted.

If you're not sure of a result, repeat the work. If you don't have the time or resources to do so, drop the whole thing or pass it on to someone else. If your facts are solid, but you're not sure of the conclusion, just say as much. If you do not have the opportunity or resources to repeat and confirm your work, don't be afraid to use words denoting timid uncertainty: "apparently," "seemingly," "suggests," "could possibly be," "raises possibility of," "may well be." If the result is worthwhile, others will either confirm or disprove what you think you found, and all will share credit. That's not sloppiness. It's just good professional conduct, true to the core of the scientific method.

Finally, remember that you enter a career in science above all in the pursuit of truth. Your legacy will be the increase and wise use of new, verifiable knowledge, of information that can be tested and integrated into the remainder of science. Such knowledge can never be harmful by itself, but as history has so relentlessly demonstrated, the way it is twisted can be harmful, and if such knowledge is applied by ideologues, it can be deadly. Be an activist as you deem necessary—and you can be highly effective with what you know—but never betray the trust that membership in the scientific enterprise has conferred upon you.

Acknowledgments

As in many of my earlier books, I am happy to acknowledge with gratitude the guidance and encouragement of my literary agent, John Taylor Williams, and my editor, Robert Weil. I also wish to acknowledge the expertise and essential dedicated hard work of my assistant, Kathleen M. Horton.

Photograph Credits

Frontispiece: Photograph by Alex Harris.

Page 12: Photograph by Howard I. Spero.

Page 20: *Boy Scout Handbook*, 4th ed. (1940), p. 643, Zoology badge emblem.

Page 26: © Paul Wiegert.

Page 42: Painting by Dana Berry of the Space Telescope Science Institute.

Page 54: English Heritage Images.

Page 68: Modified from "The political blogosphere and the 2004 U.S. election: divided they blog," by Lada A. Adamic and Natalie Glance, *Proceedings of the 3rd International Workshop on Link Discovery* (LinkKDD'05) 1: 36–43 (2005).

Page 76: Drawing by Tom Prentiss. Modified from "Pheromones," by Edward O. Wilson, *Scientific American* 208(5): 110–104 (May 1963).

Page 88: Photograph by John Hoyle.

Page 94: © Piotr Naskrecki.

Page 100: Photograph by NASA/JPL-CALTECH/ASU/UA.

Page 106: © Brian Kobilka.

Page 118: Collected by Stefan Cover in Peru. Imaged by Christian Rabeling.

Page 126: Barrett Klein, Biology Department, University of Wisconsin–La Crosse (www.pupating.org).

Page 142: W. Ford Doolittle, "Phylogenetic classification and the universal tree," figure 3, *Science* 284: 2127 (June 25, 1999).

Page 148: Catherine E. Wagner, Luke J. Harmon, and Ole Seehausen, *Nature* 487: 366–369 (2012). doi:10.1038/nature11144.

Page 168: © Klaus Bolte.

Page 176: © Abigail Lingford.

Page 188: Drawing by Tom Prentiss (moths) and Dan Todd (active space of gyplure © *Scientific American*). Modified from "Pheromones," by Edward O. Wilson, *Scientific American* 208(5): 100–114 (May 1968).

Page 204: Michael R Long © Natural History Museum Picture Library, London.

Page 218: Photograph by Daniel Simberloff.

Page 236: U.S. National Medal of Science. Government property in the public domain.

About the Author

Edward Osborne Wilson is generally recognized as one of the leading biologists in the world. He is acknowledged as the creator of two scientific disciplines (island biogeography and sociobiology), three unifying concepts for science and the humanities jointly (biophilia, biodiversity studies, and consilience), and one major technological advance in the study of global biodiversity (the Encyclopedia of Life). Among more than one hundred awards he has received worldwide are the U.S. National Medal of Science, the Crafoord Prize (equivalent of the Nobel, for ecology) of the Royal Swedish Academy of Sciences, the International Prize of Biology of Japan, and, in letters, two Pulitzer Prizes in nonfiction, the Nonino and Serono Prizes of Italy, and the International Cosmos Prize of Japan. He is currently Honorary Curator in Entomology and University Research Professor Emeritus, Harvard University.